NEUROMORPHIC OLFACTION

FRONTIERS IN NEUROENGINEERING

Series Editor
Sidney A. Simon, Ph.D.

Published Titles

Neuromorphic Olfaction
Krishna C. Persaud, The University of Manchester, Manchester, UK
Santiago Marco, University of Barcelona, Barcelona, Spain
Agustín Gutiérrez-Gálvez, University of Barcelona, Barcelona, Spain

Indwelling Neural Implants: Strategies for Contending with the *In Vivo* Environment
William M. Reichert, Ph.D., Duke University, Durham, North Carolina

Electrochemical Methods for Neuroscience
Adrian C. Michael, University of Pittsburg, Pennsylvania
Laura Borland, Booz Allen Hamilton, Inc., Joppa, Maryland

Forthcoming Titles

Statistical Techniques for Neuroscientists
Truong Young, Ph.D., The University of North Carolina at Chapel Hill

Methods in Brain Connectivity Inference through Multivariates Time Series Analysis
Koichi Sameshima, University of São Paulo, São Paulo, Brazil
Luiz Antonio Baccala, University São Paulo, São Paulo, Brazil

Humanoid Robotics and Neuroscience: Science, Engineering, and Society
Gordon Cheng, ATR Computational Neuroscience Labs, Kyoto, Japan

NEUROMORPHIC OLFACTION

Edited by
Krishna C. Persaud
University of Manchester, UK

Santiago Marco
Institute for Bioengineering of Catalonia
University of Barcelona, Spain

Agustín Gutiérrez-Gálvez
Institute for Bioengineering of Catalonia
University of Barcelona, Spain

CRC Press
Taylor & Francis Group
Boca Raton London New York

CRC Press is an imprint of the
Taylor & Francis Group, an **informa** business

CRC Press
Taylor & Francis Group
6000 Broken Sound Parkway NW, Suite 300
Boca Raton, FL 33487-2742

First issued in paperback 2019

© 2013 by Taylor & Francis Group, LLC
CRC Press is an imprint of Taylor & Francis Group, an Informa business

No claim to original U.S. Government works

ISBN-13: 978-1-4398-7171-3 (hbk)
ISBN-13: 978-0-367-38015-1 (pbk)

Visit the Taylor & Francis Web site at
http://www.taylorandfrancis.com

and the CRC Press Web site at
http://www.crcpress.com

Contents

Series Preface

The Frontiers in Neuroengineering series presents the insights of experts on emerging experimental techniques and theoretical concepts that are or will be at the vanguard of neuroscience. Books in the series cover topics ranging from electrode stability for neural ensemble recordings in behaving animals to biological sensors. In addition, there are books on statistical methods used in analyzing neuronal information. The series also covers new and exciting multidisciplinary areas of brain research, such as computational neuroscience and neuroengineering, and describes breakthroughs in biomedical engineering. The goal for this series is a reference that every neuroscientist uses to become acquainted with new advances in brain research.

Each book is edited by an expert and consists of chapters written by the leaders in a particular field. Books are richly illustrated and contain comprehensive bibliographies. Chapters provide substantial background material relevant to the particular subject.

We hope that, as the volumes become available, my effort as well as those of the publisher, the book editors, and the individual authors will contribute to the further development of brain research. The extent to which we achieve this goal will be determined by the utility of these books.

Sidney A. Simon, PhD
Series Editor

Preface

The sense of smell is highly important to survival in many animals. Changes in odorous environmental conditions cause most species (ranging from lower invertebrates to mammals) to demonstrate highly adaptive behavioral performances. Natural chemical signals such as food scents or pheromones are particularly unstable and fluctuate in quality, space, and time. Despite this, behavioral responses related to biologically meaningful odor signals can be observed even in complex natural odorous environments where backgrounds may vary enormously, demonstrating that the underlying olfactory neural network is a very dynamic pattern recognition device. We have made significant progress in understanding olfactory receptor transduction mechanisms in biological systems. The chain of events linking the presence of odorant molecules in the air (input) to the generation of a neural spike train (output) has been studied in detail, using biochemistry, molecular biology, and neurophysiology. The chain has been shown to involve odorant binding proteins, odorant degrading enzymes, receptors, G-proteins, effector enzymes, second messengers, and several ionic channels. The key step of the entire transduction process is the interaction of odorant molecules with receptor proteins—each olfactory neuron expressing a single type of receptor. The number of functional receptor types varies depending on the animal species. Odorant molecules interact with several given receptor types with specific affinities. The differential interaction of the population of receptors in the olfactory epithelium finally generates a pattern of active and inactive olfactory receptor neurons, a sensory image, which uniquely codes for the three characteristics of any odorant (or mixture of odorants), its quality, intensity (concentration), and temporal organization. One challenge is to understand how these different aspects are encoded and, on this basis, to reconstruct the global input of the olfactory population to the brain under natural and experimental conditions.

The development of artificial devices for large-scale analysis of chemical environments (such as in the "artificial olfaction" field) suffers from serious drawbacks, at the level of the gas sensors (e.g., nonselectivity, drift) and at the level of the signal processing. Nonselectivity of olfactory receptors is also encountered in biological olfaction, and it might be beneficial to look at the way biological systems process olfactory information, especially because the similarities between early olfactory systems across species imply that nature has found an optimal solution for discriminating odors. Primary olfactory centers in insects and vertebrates show striking similarities both in their cellular organization and in their types of olfactory information coding. Neurophysiological studies have revealed that neuronal circuits in the higher processing areas perform pattern decorrelation, and gain control of population activity, noise reduction, and multiplexing of different information by partial neuronal synchronization.

The book represents a concerted effort by a group of European researchers working together to unravel and translate aspects of olfaction into computationally practical algorithms that may be used in many ways to understand the underlying behavior of the chemical senses in biological systems, as well as being translated into practical applications, such as robotic navigation and systems for uniquely detecting chemical species in a complex background. The material presented is cross-disciplinary and may attract readers with a biological background or an engineering background. Neuromorphic olfaction is developed from conceptual points of view to practical applications in this book. Chapter 1 considers the biological components of vertebrate and invertebrate chemical sensing systems, which are now becoming well understood in terms of their structure and function. Artificial olfaction technology is explained, and multivariate data processing is explained conceptually, while considering biomimetic models of olfaction and how they may be translated into functional engineering devices. Chapter 2 considers the early coding pathways in the biological olfactory system showing how nonspecific receptor populations may have significant advantages in encoding odor intensity as well as odor identity. In Chapter 3 we consider the fact that in the biological system the huge redundancy and the massive convergence of the olfactory receptor neurons to the olfactory bulb are thought to enhance the sensitivity and selectivity of the system. To explore this concept, a modular expandable polymeric chemical sensor array consisting of 16,000 sensors with tens of different types of sensing materials was built and characterized. It is shown that this system has very interesting capabilities in detecting chemicals in a background, as well as discriminating mixtures of chemicals. We turn to a synthetic moth in Chapter 4 showing a neuromorphic approach toward artificial olfaction in robots. This is developed in Chapter 5, where reactive and cognitive search strategies for olfactory robots are discussed. We then consider higher animals, and a computational model of the mammalian olfactory system is discussed and implemented in Chapter 6.

This book represents a considered approach to neuromorphic olfaction that could not have happened without funding. We are grateful to the European Community for supporting our project, Biologically Inspired Computation for Chemical Sensing (NEUROCHEM) (FP7-ICT FET Project 216916).

Krishna C. Persaud
Santiago Marco
Agustín Gutiérrez-Gálvez

About the Editors

Krishna C. Persaud, FRSC, FInstMC, graduated with a BSc Honors degree in biochemistry from the University of Newcastle-upon-Tyne, UK, in 1976 and then graduated with an MSc in molecular enzymology from the University of Warwick, UK, in 1977 and a PhD specializing in olfactory biochemistry in 1980. He subsequently worked at the University of Newcastle-upon-Tyne, University of Pisa, and the Medical College of Virginia, extending his knowledge in the chemical senses. Dr. Persaud was appointed lecturer in instrumentation and analytical science at University of Manchester Institute of Science and Technology, Manchester, UK, in 1988, and progressed to his current position of professor of chemoreception in the School of Chemical Engineering and Analytical Science at the University of Manchester. He has been involved in research of chemoreception, crossing disciplines from biological aspects of olfaction to sensor arrays, electronics, signal processing and pattern recognition, and commercial development of artificial olfaction technologies.

Santiago Marco completed his licenciatura degree in applied physics in 1988 and received a PhD in microsystem technology from the University of Barcelona in 1993. He held a European Human Capital Mobility grant for a postdoctoral position in the Department of Electronic Engineering at the University of Rome "Tor Vergata." Since 1995, he has been an associate professor of electronic instrumentation in the Department of Electronics at the University of Barcelona. In 2004 Dr. Marco spent a sabbatical leave at EADS Corporate Research, Munich, to work on ion mobility spectrometry. In 2008 he was appointed leader of the Artificial Olfaction Lab at the Institute of Bioengineering of Catalonia. Dr. Marco's research concerns the development of signal/data processing algorithmic solutions for smart chemical sensing based in sensor arrays or microspectrometers integrated typically using microsystem technologies.

Agustín Gutiérrez-Gálvez received his BE degrees in physics and electrical engineering from the University of Barcelona in 1995 and 2000, respectively. He received his PhD degree in computer science from Texas A&M University in 2005. In 2006 Dr. Gutiérrez-Gálvez was a JSPS postdoctoral fellow at Tokyo Institute of Technology and came back to the University of Barcelona with a Marie Curie fellowship. Currently, he is an assistant professor in the Department of Electronics at the University of Barcelona. His research interests include biologically inspired processing for gas sensor arrays, computational models of the olfactory system, pattern recognition, and dynamical systems.

Contributors

Romeo Beccherelli
Institute for Microelectronics and
  Microsystems
Consiglio Nazionale delle Ricerche
  (CNR-IMM)
Rome, Italy

Simon Benjaminsson
Department of Computational Biology
Royal Institute of Technology
and
Stockholm Brain Institute
Stockholm, Sweden

Sergi Bermúdez i Badia
Laboratory for Synthetic Perceptive,
  Emotive, and Cognitive Systems
Universitat Pompeu Fabra (UPF)
Barcelona, Spain

Mara Bernabei
School of Chemical Engineering and
  Analytical Science
University of Manchester
Manchester, United Kingdom

Alex Escuredo Chimeno
Laboratory for Synthetic Perceptive,
  Emotive, and Cognitive Systems
Universitat Pompeu Fabra (UPF)
Barcelona, Spain

Agustín Gutiérrez-Gálvez
Intelligent Signal Processing Group
Department of Electronics
University of Barcelona
and
Artificial Olfaction Laboratory
Institute for Bioengineering of
  Catalonia
Barcelona, Spain

Pawel Herman
Department of Computational Biology
Stockholm University
and
Stockholm Brain Institute
Stockholm, Sweden

Anders Lansner
Department of Computational Biology
Royal Institute of Technology
and
Stockholm Brain Institute
Stockholm, Sweden

Lucas L. Lopez-Serrano
Laboratory for Synthetic Perceptive,
  Emotive, and Cognitive Systems
Universitat Pompeu Fabra (UPF)
Barcelona, Spain

Santiago Marco
Intelligent Signal Processing Group
Department of Electronics
University of Barcelona
and
Artificial Olfaction Laboratory
Institute for Bioengineering of
  Catalonia
Barcelona, Spain

Dominique Martinez
Laboratoire Lorrain de Recherche en
  Informatique et ses Applications
  (LORIA)
Centre National de la Recherche
  Scientifique (CNRS)
Vandoeuvre-lès-Nancy, France

Eduardo Martin Moraud
Laboratoire Lorrain de Recherche en
    Informatique et ses Applications
    (LORIA)
Centre National de la Recherche
    Scientifique (CNRS)
Vandoeuvre-lès-Nancy, France

Zenon Mathews
Laboratory for Synthetic Perceptive,
    Emotive, and Cognitive Systems
Universitat Pompeu Fabra (UPF)
Barcelona, Spain

Simone Pantalei
Institute for Microelectronics and
    Microsystems
Consiglio Nazionale delle Ricerche
    (CNR-IMM)
Rome, Italy

Alexandre Perera i Lluna
Research Centre for Biomedical
    Engineering
Universitat Politecnica de Catalunya
    (UPC)
Barcelona, Spain

Krishna C. Persaud
School of Chemical Engineering and
    Analytical Science
University of Manchester
Manchester, United Kingdom

Paul F. M. J. Verschure
Laboratory for Synthetic Perceptive,
    Emotive, and Cognitive Systems
Universitat Pompeu Fabra (UPF)
and
Institucio Catalana de Recerca i Estudis
    Avancats (ICREA)
Barcelona, Spain

Vasiliki Vouloutsi
Laboratory for Synthetic Perceptive,
    Emotive, and Cognitive Systems
Universitat Pompeu Fabra (UPF)
Barcelona, Spain

Emiliano Zampetti
Institute for Microelectronics and
    Microsystems
Consiglio Nazionale delle Ricerche
    (CNR-IMM)
Rome, Italy

Andrey Ziyatdinov
Research Centre for Biomedical
    Engineering
Universitat Politecnica de Catalunya
    (UPC)
Barcelona, Spain

1 Engineering Aspects of Olfaction

Krishna C. Persaud

CONTENTS

ABSTRACT

The biological components of vertebrate and invertebrate chemical sensing systems are now becoming well understood in terms of their structure and function. As a result, it is now possible to consider biomimetic models of olfaction and how they may be translated into functional engineering devices. This chapter outlines some of the development in sensor array technology that can enable functional devices. The responses of individual odor sensors combined into an array, where each sensor possesses slightly different response selectivity and sensitivity toward the sample odors, when combined by suitable mathematical methods, can provide information to discriminate between many sample odors. Arrays of gas and odor sensors, made using different technologies, have become known as electronic noses and consist of three elements: a sensor array that is exposed to the volatiles, conversion of the sensor signals to a readable format, and software analysis of the data to produce characteristic outputs related to the odor encountered. The output from the sensor array may be interpreted via a variety of methods—such as pattern recognition algorithms, principal component analysis, discriminant function analysis, cluster analysis, and artificial neural networks—to discriminate between samples. This chapter introduces some of the more biologically oriented algorithms that may transform traditional multivariate data analysis, feature extraction, and pattern recognition.

1.1   INTRODUCTION

Animals have evolved an incredible diversity of sensory systems to extract information from the environment. Of these, the chemosensory systems allow them to extract information from their chemical environments, so that behavioral preferences are elicited in response to stimuli that may be aversive or attractive. Animals live in complex environments where an infinite variety of chemical molecules may be encountered. These may be present as single chemicals, or as complex mixtures, where the relative concentrations of individual components differ. The tasks commonly carried out by the olfactory system include detection of odors, estimating their strength, identifying their source, and recognizing a specific odor in the background of another. The olfactory system in mammals is involved in physiological regulation, emotional responses (e.g., anxiety, fear, pleasure), reproductive functions (e.g., sexual and maternal behaviors), and social behaviors (e.g., recognition of members of the same species, family, clan, or outsiders). In insects such as the honeybee, it has been shown that scents modify behaviors associated with mating, foraging, recognition of kin, brood care, swarming, alarm, and defense (Reinhard and Srinivasanand 2009).

Figure 1.1 shows a diagram of the olfactory epithelium of a mammal. Olfactory receptor neurons are bipolar, and from the apical side, cilia containing membrane-bound olfactory receptor proteins lie in an aqueous environment (mucus) overlying the epithelium. Odorant molecules need to partition from air into water before they can reach the transduction sites in the epithelium. Soluble odorant binding proteins are secreted into the aqueous mucus layer, and these may have an odorant carrier and preconcentration role. Over the last century, ideas that several classes of olfactory receptors exist, selective to chemical species on the basis of molecular size, shape, and charge, were based on evidence from chemistry (Beets 1978), olfactory psychophysics, and structure-activity relationships of odorants (Boelens 1974), together with the examination of "specific anosmias" in the human population, which all supported the definition of selectivity and specificity of putative olfactory receptors initiated by Amoore (1962a, 1962b, 1967). These were confirmed by developments in olfactory neurobiology and molecular genetics (Buck and Axel 1991; Buck 1997a, 1997b; Chess et al. 1992; Mombaerts et al. 1996a). The ideas that several classes of olfactory receptors exist, selective to chemical species on the basis of molecular size, shape, and charge, also pointed to individual olfactory receptors being rather broad in their selectivity to molecules within certain classes. The important molecular parameters

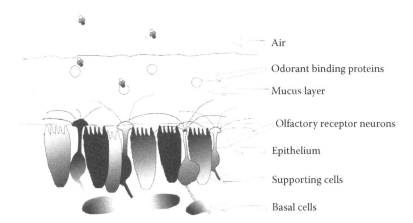

FIGURE 1.1 Olfactory epithelium of vertebrates. The interface between air and the olfactory receptors is a layer of mucus, where soluble odorant binding proteins are found. These may have an odorant carrier function. The olfactory cells of the epithelium are bipolar neurons that congregate to form the olfactory nerve. The apical side of these neurons is covered with nonmotile cilia—with the plasma membrane containing olfactory receptor proteins. On the distal side of the epithelium basal cells develop into new olfactory neurons—there is constant turnover of olfactory receptor neurons.

of an odorant determining the olfactory response would include the adsorption and desorption energies of the molecule from air to a receptor interface, partition coefficients, and electron donor-acceptor interactions, depending on the polarizability of the molecule, and its molecular size and shape. The plethora of chemicals that an animal can sense, as well as their combinatorial and temporal variability, has made it difficult to understand how the brain processes the incoming information so that an animal can make sense of its chemical environment. Polak (1973) proposed a multiple profile–multiple receptor site model for vertebrate olfaction anticipating some of the combinatorial coding mechanisms later discovered. The identification of odorant receptor (OR) genes in rodents (Buck and Axel 1991), in *Caenorhabditis elegans* (Sengupta et al. 1996), and in *Drosophila melanogaster* (Clyne et al. 1999; Gao and Chess 1999) have given us a fundamental understanding of olfactory coding, especially at the olfactory receptor neuron (ORN) level. Individual ORs are proteins that traverse the cell membrane of the cilia of the olfactory neuron. It appears that there may be hundreds of odorant receptors, but only one (or at most a few) expressed in each olfactory receptor neuron. These families of proteins may be encoded by as many as 1000 different genes in humans. This is a large number and accounts for about 2% of the human genome. In humans, however, most are inactive pseudogenes, and only around 350 code for functional receptors. There are many more functional genes in macrosmatic animals like rats. These receptor proteins are members of a well-known receptor family called the seven-transmembrane domain G-protein-coupled receptors (GPCRs) (see Figure 1.2). The hydrophobic regions (the transmembrane parts) contain maximum sequence homology to other members of the G-protein-linked receptor family. There are some notable features of these olfactory receptors, like the divergence in sequence in the third, fourth, and fifth

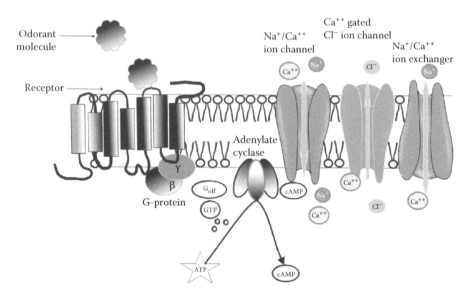

FIGURE 1.2 Olfactory receptor and transduction mechanism. Odorants in the mucus bind directly (or are transported via odorant binding proteins) to receptor molecules located in the membranes of the cilia. The conformational change occurring on binding activates an odorant-specific G-protein (G_{olf}) that in turn activates an adenylate cyclase, resulting in the generation of cyclic AMP (cAMP). A target of cAMP is a cation-selective channel that allows the influx of Na^+ and Ca^{2+} into the cilia. The rise in intracellular Ca^{2+} allows opening of Ca^{2+}-gated Cl^- channels that cause depolarization of the olfactory receptor potential. Phosphodiesterases inactivate cAMP, and this allows return of the receptor potential to its resting state. At the same time, Ca^{2+} complexes with calmodulin (Ca^{2+}-CAM) and binds to the channel, reducing its affinity for cAMP. Finally, Ca^{2+} is eliminated through the $Ca^{2+}/Na+$ exchange pathway.

transmembrane domains, that suggest how a large number of different odorants may be discriminated (Pilpel and Lancet 1999).

As crystallographic information on olfactory receptors is lacking, they have been modeled based on their resemblance to rhodopsin. Gelis et al. (2012) has published models of putative binding sites of some human olfactory receptors. On the inner side of the cell membrane, proteins called G-proteins are associated with olfactory receptor. These bind the guanine nucleotides—guanine triphosphate and guanine diphosphate. They are made up of three subunits and are located with the inner surface of the plasma membrane. They are closely associated with the transmembrane receptor protein. When an odorant binds, it is thought that an allosteric change in conformation occurs, in turn causing a conformational change in a subunit of the G-protein G_α—displacing bound guanine triphosphate (GTP) and allowing it to bind GTP. This in turn produces an activated subunit that dissociates from the other subunits and activates another effector molecule, triggering a cascade of events that leads to the opening of an ion channel, and change of electrical potential across the cell membrane. As this electrical potential propagates to the basal side of the cell, it triggers in turn voltage-gated ion channels so that a series of electrical spikes results, which are transmitted to the processing centers in the brain via the axon of the olfactory neuron.

Our understanding is that mammalian and insect olfactory systems are combinatorial in nature—instead of activating a single specialized receptor, each chemical stimulus induces a complex pattern of responses across the olfactory receptor array. The investigation of OR expression patterns has made it possible to dissect the major circuits underlying olfaction (Hoare et al. 2011; Imai et al. 2010; Leinwand and Chalasani 2011; Ressler 1994; Ressler et al. 1993; Su et al. 2009; Vassar et al. 1993). The evidence obtained confirmed previous concepts of a common design of mammalian and insect olfactory systems that are discussed by Hildebrand and coworkers (Hildebrand and Shepherd 1997; Hildebrand 2001; Martin et al. 2011). The consequence of the combinatorial design of the olfactory system is that the number of unique odor representations is not limited to the number of different receptor types, but can be estimated as m^n, where n corresponds to the number of receptor types available and m the number of possible response states that each sensor can assume. This is limited to the available signal-to-noise parameters associated with the working system (Cleland and Linster 2005).

Vertebrate or invertebrate life surviving in complex, changing environments requires the use of sophisticated sensory systems to detect, classify, and interpret patterns of input stimulation. Coding mechanisms by which a certain pattern of stimulations may be described are inherent. Such codes may be defined as sets of symbols that can be used to represent patterns of organizations and the sets of rules that govern the selection and use of these symbols. Sensory coding mechanisms in biological systems would appear to project some representation of sensory inputs as a pattern at a high level of the nervous system, the neural activity resulting being then related to the previous experience with regard to this pattern or associated patterns. Fundamental concepts of pattern classification that seem to be common in biological systems would appear to be *template matching*, whereby the pattern to be classified is compared with a set of templates, one for each class, the closest match determining the classification, and *feature detection systems*, in which a number of measurements are taken on the input pattern and the resulting data are combined to reach a decision. These systems may involve either a sequential approach whereby information from the evaluation of some features is used to decide which features to evaluate next, or a parallel approach where information about all features is evaluated at the same time with no weight being placed on any particular feature.

The remarkable capabilities of the biological chemosensory systems in detection, recognition, and discrimination of complex mixtures of chemicals, together with rapid advances in understanding how these systems operate, have stimulated the imagination and interest of many researchers and commercial organizations for the development of electronic analogs. The dream of emulating biological olfaction using artificial devices was conceptually realized by Persaud and Dodd (1982), who demonstrated that an array of electronic chemical sensors with partial specificity could be used to discriminate between simple and complex odors; i.e., the combinatorial aspects of olfactory receptors could be emulated, and this could achieve remarkable flexibility in terms of the numbers of types of analytes that can be discriminated. This led to a burgeoning of the "electronic nose" field of research, and formation of many commercial enterprises interested in exploiting a wide range of applications, including environmental, food, medical, security, and others. The researchers and companies have produced instruments that combine gas sensor arrays and pattern analysis techniques for the detection, identification, or quantification of volatile compounds. The multivariate response of an array of chemical gas sensors with broad and partially overlapping selectivities can be processed as a pattern or "fingerprint" to discriminate a wide range of odors or volatile compounds using pattern recognition algorithms. The instruments typically consist of a gas sensor array comprising many types of sensing technologies, a sample delivery system, and the appropriate electronics for signal processing, data acquisition, and storage. Processing of data from such systems can be split into four sequential stages: signal preprocessing, dimensionality reduction, prediction, and validation. The numbers of sensors incorporated into the devices are relatively small, and the data handling approaches have been based on traditional chemometric or neural network methods for processing multivariate data. Applications using such chemosensory arrays at present involve issues such as sensor drift, poor sensitivity compared to biological systems, and interference from background odors. With further understanding of biological processes, some of these engineering limitations may be reduced by the adaptation of biologically plausible models for signal processing.

This chapter gives an introduction to biological chemoreception, going on to the field of artificial olfaction, and discussing some of the signal processing concepts that may be useful in mimicking biological olfactory systems.

1.2   ODORANTS, CHEMICAL STIMULI, AND ODORANT BINDING SITES

Odorants are relatively small molecules of the order 10^{-8} to 10^{-10}m in size. There are many aspects, such as shape and charge distribution, that determine a molecule's properties and in turn whether they exist as odorants and their particular scent. Odorants that are typically found in nature tend to be a vast complex mixture of many different chemical species at various concentrations rather than one type of molecule. Varying the concentration of compounds in the aromatic mixture, changing the functional group chemistry or position, and altering the stereochemistry or conformation of a molecule, among other changes that can be made, all affect the smell of an odorant or odorous mixture (Beets 1978). Multiple molecular properties of the stimulus molecules, including these individual molecular properties or determinants, population properties of a single odor substance, and appropriate mixtures of different odor substances, all affect the perception of a volatile chemical stimulus (Shepherd 1990). Although there are examples of simple single molecules with distinct aromas, the discovery of the olfactory code is an extremely complicated task. Odor molecules may vary in structure from simple aliphatic short-chain organic molecules to complex aromatic molecules with multiple functional groups, as shown in Figure 1.3. It illustrates that small changes in the nature of the functional group, the shape, and the size of a molecule can induce great changes in our perception of odor quality (how we describe our perception of an odor). In many cases a change of a single carbon atom can give rise to different odor qualities in terms of human

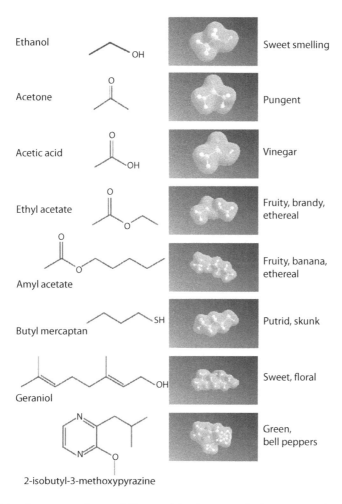

Ethanol — Sweet smelling

Acetone — Pungent

Acetic acid — Vinegar

Ethyl acetate — Fruity, brandy, ethereal

Amyl acetate — Fruity, banana, ethereal

Butyl mercaptan — Putrid, skunk

Geraniol — Sweet, floral

2-isobutyl-3-methoxypyrazine — Green, bell peppers

FIGURE 1.3 Structure and space filling models of a range of odorants. The size, shape, and functional groups determine the quality of the odorant molecules perceived.

perception, while in other cases apparently unrelated chemicals produce similar odor perceptions, as shown in the case of almond odor in Figure 1.4, where it can also be seen that the shapes of these chemically different molecules may be very similar.

Odorants must be volatile in order to be able to be carried to the nasal cavity. Therefore, the molecular weight is confined to be around 30 to 300 g.mol^{-1}. Other important factors are charge distribution and polarizability. Nonpolar and symmetrical molecules tend to be gaseous at room temperature, but not odorants. Introduction of one or two functional groups establishes some polarity and better interaction between molecules. Hydrogen bonding is also expected to play an important role in odorants, particularly the interaction of the odorant molecules and the olfactory receptor proteins. These characteristics are typical of odorants and relate to their affinity and

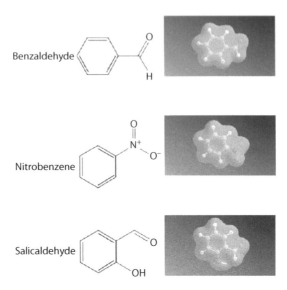

Benzaldehyde

Nitrobenzene

Salicaldehyde

FIGURE 1.4   Different chemicals may produce similar odor perceptions: changing the functional groups on the ring does not change the perception of almond odor from benzaldehyde, nitrobenzene, and salicaldehyde.

interaction with the olfactory receptors. The recognition of an odor also depends on the threshold detection limit, which varies between different people and greatly varies between different animals.

Molecules containing covalently bonded sets of atoms with a delocalized conjugated system of π-orbitals in a coplanar arrangement with one or more rings have a special place in organic chemistry. If the number of delocalized π-electrons is even but not a multiple of 4 (Hückel's rule), the molecule is called aromatic. Although this term has a precise definition, the name arose from the fact that many of these types of molecules have a distinctive odor.

It is important to understand what properties of a molecule are mapped by the olfactory system. These include the presence of functional groups, geometry (e.g., molecular length, position of functional groups, geometry of double bonds), and connectivity (e.g., number and sizes of rings and branching. Weyerstahl (1994) reviewed structure-odor correlations, and some of the concepts described have been developed further and updated by Sell (2006). Many odorants contain only one strongly polar function in the molecular structure, and this polar group is thought to form a hydrogen bond or some other dipolar attachment to a polar site on an olfactory receptor, with the remainder of the molecule occupying a hydrophobic space in the receptor. The polar groups such as aldehydes, ketones, esters, ethers, and also some heteroatoms, such as nitrogen and sulfur, are called osmophores (Rupe and von Majewski 1900), and they are used as a reference point in the molecule when mapping structure-activity relationships (SARs) as molecular reference points. A second, usually weaker, electron donor or acceptor may also sometimes be involved.

Relating the structure of an odorant molecule to the sensation that is perceived is not an easy task, and there are many unexplained findings in work so far published.

The interaction of a molecule with an olfactory receptor translates into a mixture of repulsive and attractive forces that a molecule encounters when bound to a receptor, and it is clear that simplistic shape models of odorant-receptor interaction do not quite account for the observations. Different types of molecules can have similar odors, while similar molecules can have dissimilar odors; e.g., a series of molecules of different chemical families all have an almond odor to humans (Figure 1.4), while two very similar molecules, such as D-carvone and L-carvone, or vanillin and iso-vanillin (Figure 1.5), elicit very different sensory perceptions. These are discussed in some detail by Turin and Yoshii (2003) in terms of possible mechanisms that are

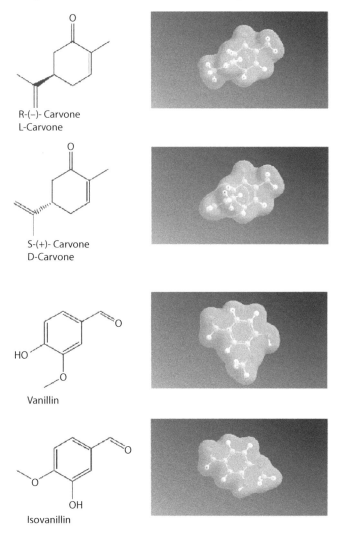

FIGURE 1.5 Similar molecules may produce different sensory perceptions. Isomers of carvone have different odor qualities: L-carvone, spearmint; D-carvone, caraway. Similarly, vanillin and isovanillin have different odors—vanilla and a weak phenolic odor, respectively.

involved in transduction of these molecules. Some olfactory receptors appear to be tuned to certain families of chemicals; e.g., when the receptive range of OR-I7 was mapped (Araneda et al. 2000), this indicated good selectivity to a range of aliphatic aldehydes up to chain size C11. New methods allowing expression of olfactory receptors on the surface of human embryo kidney cells (HEK293), which are then loaded with fluorescent dyes allowing calcium imaging to be carried out, have given us a new tool to map the selectivity of olfactory receptors (Sanz et al. 2005). Unlike the OR-I7 receptor, human OR-1G1 receptor seems to be quite selective within a class of substrates (for example, alcohols or acids), but not selective between classes, as it responds to alcohols, aldehydes, acids, esters, lactones, and a variety of heterocyclic systems (Sanz et al. 2005). Spehr and Leinders-Zufall (2005) showed that activation of human olfactory receptor (hOR-17) by bourgeonal is inhibited by undecanal, a nonagonist, suggesting allosteric interactions and hence multiple binding sites. Triller et al. (2008) suggest that receptor activation rather than odor should be used as input data in order to understand odorant-receptor interaction.

Objective description of odor perception in humans is a complex task. Methods used include

1. The use of profiles of reference odors where an odorous substance is described in terms of a similarity profile that is related to a certain number of reference substances covering the olfactory space.
2. The use of profiles of semantic descriptors: An odorous substance is described by means of a list of semantic descriptors, the values of which can vary in intensity.
3. The use of odor similarities: The likeness between two substances is ranked on a numerical scale fixed a priori (Callegari et al. 1997). They deduce from an analysis of published data that the estimation of a similarity between odors is based only on the distinctive elements; the common elements do not add any information. They also indicate that 25 well-chosen descriptors are enough to describe the perceptual olfactory space.

To model the molecular characteristics that might account for our perceptions, quantitative structure-activity relationship (QSAR) studies are useful tools. It can be considered that a molecule makes a random walk from a very dilute solution until it hits the appropriate receptor or binding site that elicits a response. The rate at which this biological response occurs is dependent on the properties of the molecule in question.

$$\frac{dR}{dt} = AkC_x \tag{1.1}$$

where R is the response, A is the probability of a molecule reaching a binding site in a defined time interval, k_x is a rate constant or equilibrium constant, and C_x is the extracellular concentration of the molecules. The most common properties that are correlated to biological activity are electronic properties and lipophilicity. The parameters used to measure electronic and lipophilicity properties are σ, Hammett values, and π, a lipophilicity constant, respectively (Hansch and Fujita 1964). Although other parameters have been investigated in QSAR equations, σ and π are the most widely accepted. With these two parameters, a typical QSAR equation takes the form of Equation 1.2:

$$\log\left(\frac{1}{C}\right) = -k\pi + k'\pi^2 + \rho\sigma + k'' \tag{1.2}$$

Log $1/C$ is a term representing the concentration of a drug needed to achieve a desired level of effect. k, k', ρ, and k'' are all regression coefficients. The π term is present as both π and π² since lipophilicity tends to follow a parabolic relationship relative to activity. In a study of structurally similar odorless and odoriferous benzenoid musks, Yoshii and coworkers (Yoshii et al. 1991, 1992) searched for the best molecular parameters to discriminate these groups. They found that the best three parameters were the log P value (partition coefficient between octanol and water, indicating the balance of lipophilicity and hydrophilicity), the longest side length of a hexahedron circumscribing the molecule, and the parameter that expresses the structural hindrance to the functional group when a molecule approaches the receptor site.

Similar QSAR techniques have been adapted by Guo and Kim (2010) to map the electrophysiological responses of *Drosophila* to 108 odorants. It is worth dwelling on some of the important findings. *Drosophila* olfactory receptors are thought to use a combinatorial mechanism to encode olfactory information at the receptor level. Odors may either stimulate or inhibit receptor responses. It was found that linear odorants with five to eight nonhydrogen atoms at the main chain and a hydrogen bond acceptor or hydrogen bond donor at its ends will stimulate a strong excitatory response. They compared the sequences of 90 ORs in 15 orthologous groups and identified 15 putative specificity-determining residues (SDRs) and 15 globally conserved residues that were postulated as functionally key residues. On mapping these residues to models of secondary structure it was found that 12 residues were located in transmembrane domains, while others were located in extracellular halves of its transmembrane domains. As a result, it was hypothesized that the odorant binding pocket lies on this extracellular region. Evidence from QSAR modeling indicates that the binding pocket is about 15 angstroms in depth by about 6 angstroms in width, with 12 key polar or charged residues located in this pocket, and functioning in distinguishing between docked odorants on the basis of geometrical fitting of the molecule into the binding pocket, and hydrogen bond interactions. It seems that many of the original concepts of the receptor site binding outlined by Amoore (1962a) are now being verified.

Many researchers have attempted to model the binding sites for individual olfactory receptors. The QSAR approach can give valuable models of the ligand structure, and information on the form of the binding pocket. Receptor-based approaches such as homology modeling create a model of the protein and the binding site explicitly, and from this give information on ligand binding (Crasto 2009). Both techniques can be combined together. Gelis and coworkers (2012) have used a molecular dynamic approach to predict which amino acid residues play a crucial role in binding odorants. They combined dynamic homology modeling with site-directed mutagenesis and functional analysis, to produce a molecular model of the ligand binding niche of hOR-2AG1 within a receptor model. They deduced the basis for receptor activation by ligands based on computed hydrogen bond contact frequencies to amino acids forming the ligand binding site, forming a ligand selectivity filter. The information gives new insight into the interaction of volatile, highly hydrophobic, and flexible ligands with olfactory receptors.

1.3   ODOR SAMPLE

What is perceived by an animal is a stimulus that is fluctuating constantly in terms of its composition as well as instantaneous concentration. The native environment of an animal provides varied samples of odor stimuli in space and time (see Figure 1.6). It is thought that insect olfactory systems work as flux detectors rather than concentration analyzers, since they react rapidly to changes in the local environment (Kaissling and Rospars 2004; Rospars et al. 2007). Odor samples that are being perceived consist of short bursts where the instantaneous concentration may vary widely. Odor plumes are often meandering and subject to molecular and turbulent diffusion. When near to a source, the peak-to-mean ratio is smaller, and when farther away from the source, the bursts are on average weaker but longer and with greater gaps between them, but there are always exceptions to this. Insects are capable of resolving bursts of odor at 10 s^{-1}. Murlis and coworkers (Murlis et al. 1992; Murlis and Jones 1981) describe mathematical models that allow a description of the wisps of odor that are emitted from a source. A plume could be described in terms of large-scale structure, where the shape and average odor strength help an animal to

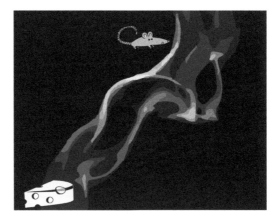

FIGURE 1.6   Odor plumes. An odor plume can be thought of as being sheared off as a strong single strand from the source of emission. Air turbulence shreds the odor plume into many substrands as it is transported by larger-scale turbulence out into the environment. These substrands may contain highly concentrated odor. There is a large-scale structure that may be perceived—the shape and average structure of the plume, small-scale structure due to the fluctuating substrands, and a time-averaged structure that determines how the biological nose may contact the plume at various points downwind from the source.

orient, or in more detail, in terms of the small-scale structure within the plume itself, where fluctuating plume concentrations influence the behavior of an animal when it is within the plume, or in terms of a time-averaged structure where the animal may contact a plume at different points downwind from the source.

As a rat sniffs an odor, the air in the nasal passages is conditioned and changes in temperature and humidity, so that by the time molecules arrive at the olfactory epithelium, the air is at fairly constant temperature and humidity. Sniffing dynamics may also direct the stimulus to optimal portions of the sensory epithelium (Schoenfeld and Cleland 2006), and head and body movements help an animal detect and locate an odor. Once an odor is detected, attentional behavior associated with any sensory stimulus will occur. However, we know relatively little about the dynamical interaction between brains and their environments.

1.4 OLFACTORY SYSTEMS

Insect and vertebrate olfactory systems show many analogies in terms of organization (Hildebrand and Shepherd 1997), as shown in Figure 1.7. It is thought that these olfactory systems evolved independently or that there was a common olfactory ancestor with subsequent drastic divergence in the development of olfactory receptors (Benton 2009), with these evolutionary sequences converging on similar types of algorithms (Ache and Young 2005; Kay and Stopfer 2006). Both vertebrate and

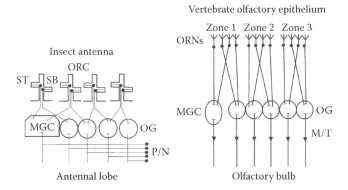

FIGURE 1.7 Comparative olfactory systems of insects and vertebrates. In insects, ORNs are located and compartmentalized in olfactory sensilla, while in vertebrates ORNs are found in the olfactory epithelium. In insects there are different types of sensilla, such as trichoid sensilla (ST) and basiconic sensilla (SB), containing different types of olfactory receptors. In both insects and vertebrates glomerular (GL) units receive input from olfactory neurons that respond to a range of odorants. Axons of pheromone-sensitive ORNs project into a group of enlarged glomeruli, called the macroglomerular complex (MGC). In vertebrates, the mitral or tufted cell (M/T) exhibits excitatory responses to a narrower range of odorants as well as inhibitory responses to molecules related to the excitatory odorants, and similar behavior is seen in the projection neurons (P/N) in insects. The narrowness of this response relative to the input to the glomerular unit is thought to result from lateral inhibitory mechanisms mediated by reciprocal dendrodendritic synapses between mitral and granule (GR) cells in vertebrates and the olfactory receptor neurons and projection neurons in insects.

invertebrate olfactory systems need to perform similar tasks. They need to detect potential signals of interest from chemically noisy environments. They have the task of feature extraction to extract these signals from a complex and changing odor background to form internal representations of the chemical stimuli, and then to compare these patterns to those of previously experienced odors. They further need to differentiate relevant from irrelevant stimuli, which may be context dependent, and then make an appropriate behavioral or other response. From anatomical, physiological, and behavioral evidence, many of the neural circuit elements comprising the olfactory system have been proposed to contribute to these processes in particular ways; for example, multiple feedback and feed-forward interactions among olfactory structures, as well as between olfactory and nonolfactory areas, are thought to contribute to the filtering and construction of olfactory representations (Ache and Young 2005).

From the periphery of the olfactory system to the higher processing areas of the brain there are several key components. In considering the design of artificial or biomimetic chemosensory systems, some of the important features are discussed in the following sections.

1.4.1 PERIPHERAL SYSTEMS

As discussed above, odorants are small molecules that must partition from air to an aqueous phase in order to reach the binding sites of the receptor cells. In vertebrates the mucus layer overlying the olfactory epithelium contains an ionic medium bathing the olfactory receptors (Figure 1.1). This medium contains small water-soluble proteins that are called odorant binding proteins because they are capable of binding several types of odorant molecules with low affinity. Similar proteins are also found in insects. In fact, olfactory binding proteins (OBPs) are small water-soluble proteins that constitute one of the largest groups of proteins involved in olfaction (Vincent et al. 2000, 2004; Xu 2005; Xu et al. 2009). In vertebrates OBPs form two groups: OBPs and pheromonal binding proteins (PDPs). Vertebrate OBPs belong to a superfamily of proteins called lipocalins, which are commonly involved in the transport of hydrophobic molecules. Insect OBPs can be divided into three groups: PDPs, general odor binding proteins (gOBPs), and antennal specific proteins (ASPs) (Pelosi and Maida 1995; Forêt and Maleszka 2006). These two groups of OBPs are not homologous and are thought to have arisen by convergent evolution (Forêt and Maleszka 2006). However, the similarities between these groups of proteins make solid distinctions and specific role allocations difficult.

OBPs require a significant amount of energy in order to maintain their very high turnover rate. This implies that they have an important physiological function. They are secreted in high concentrations by nonneuronal cells into the fluid surrounding olfactory dendrites. This aqueous fluid provides a barrier between primarily hydrophobic odorant molecules and olfactory receptors (ORs). In invertebrates this fluid is sensilla lymph, and in vertebrates it is a layer of mucus. OBPs are thought to solubilize and shuttle odorants through this aqueous fluid to allow physical contact with ORs, and this is proposed to be their primary role (Pelosi 2001). This is supported by the fact that OBPs of many species have been shown to bind reversibly and selectively to a wide range of odorants (Xu 2005; Kim et al. 1998). Besides the shuttle

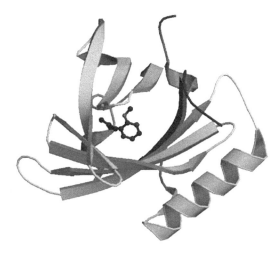

FIGURE 1.8 Porcine OBP. The secondary structure of the OBP is characterized by a central antiparallel β-barrel and an α-helix, which are characteristic of the lipocalin superfamily. A large buried cavity in the β-barrel forms the ligand binding site. The ligand isobutylmethoxypyrazine is shown in the binding pocket of the protein.

role described above, OBPs have been suggested to have many other roles, including odor recognition (Forêt and Maleszka 2006), odor release (Hekmat-Scafe et al. 2002), concentration of odorants (Pelosi and Maida 1995), and protection of nasal mucosa and odorant scavenging to prevent receptor saturation (Tegoni et al. 2000). Figure 1.8 shows a schematic diagram of porcine OBP and the ligand 2-isobutyl-3-methoxy pyrazine interacting with the binding site. The protein belongs to the lipocalin family, and is conformationally very stable.

From an engineering perspective, the function of these proteins has interesting consequences. Focusing on the pheromone binding protein (PBP), it is thought that it may have the following functions in insects:

1. Solubilization of the pheromone (which is a lipophilic molecule)—transport through the sensillar lymph, which prevents it from entering the cell membrane
2. Protection of the pheromone from enzymatic degradation
3. Pheromone receptor activation
4. Pheromone deactivation (scavenging function)
5. Provision of organic ions to the sensillar lymph

There is controversy over these functions, and Vogt and coworkers (Vogt 2006; Vogt and Riddiford 1981, 1984) differ in opinion from Kaisling (2009) over the details of these steps, but the latter group have modeled the kinetics of the various pathways and explored a variety of models that have evolved over time and have come up with some numbers that are interesting (see Figure 1.9). The consequence of the sequestration of pheromone by the pheromone binding protein is that it acts as a preconcentrator. After adsorption from the air space at the surface of the olfactory hair, the pheromone (F) passes through the hair wall via the pore tubules. The

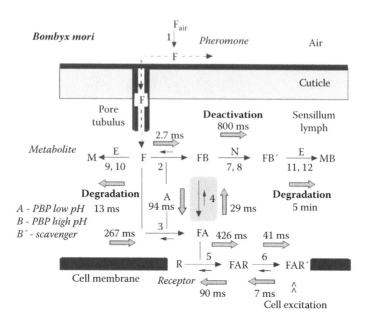

FIGURE 1.9 Temporal model characteristics (model N) by Kaissling (2009). Model of the time constants associated with PBP binding in Bombyx mori. The pheromone is adsorbed by the sensillum hair (reaction 1) and diffuses along the hair surface, through the hair wall via the pore tubules, and—carried by the pheromone binding protein—toward the receptor neuron. Entering the hair lumen the pheromone (F) binds to one of two reaction partners dissolved in the sensillum lymph—either the pheromone-degrading enzyme (E) or the A-form of the PBP (A). When F binds to the A-form at neutral pH (reaction 2), the complex FB is produced, changing the conformation of the PBP. The binding of F and A at low pH forms the complex FA (reaction 3). Upon pH changes the complex FB may be converted into FA and back to FB (reaction 4). The complex FA is assumed to be the only species binding to the receptor molecule (R) (reaction 5). The ternary complex (FAR) may go to an activated state (FAR′) (reaction 6), which initiates excitation of the neuron via rapid cellular signal processes. Binding of F and the reduced form of the binding protein forms a complex FB with a hypothetical enzyme N catalyzing the deactivation of the complex FB. Reactions 7, 9, and 11 represent binding to and dissociation from the enzymes E and N. The catalytic reactions 8, 10, and 12 are irreversible. (Reproduced from Kaissling, K. E., *Journal of Comparative Physiology A—Neuroethology Sensory Neural and Behavioral Physiology*, 195 (10), 895–922, 2009. With permission.)

pheromone is then transported to the receptor neuron while bound to the PBP. It is thought that most of the pheromone entering the hair lumen (the fraction Q1) binds to the PBP. A portion of the incoming pheromone (1-Q1) encounters a pheromone-degrading enzyme (E) within the sensillum lymph, is rapidly degraded to a metabolite (M), and no longer functions as a stimulus compound. It is assumed that the pheromone-PBP complex rather than the free pheromone interacts with the receptor molecules (R). A single activation of the pheromone-PBP-receptor complex is thought to elicit an elementary receptor potential (Minor and Kaissling 2003). From experimental data and modeling, they arrive at the temporal characteristics shown in Figure 1.9. After entering the hair lumen, the free pheromone F may bind to PBP or be degraded enzymatically. It has a half-life of about 3 ms due to binding to the PBP and the formation of the complex between the pheromone and the pheromone binding protein (reaction 2). This half-life is much shorter than expected from solely enzymatic degradation (13 ms). Taking PBP binding and degradation together, the overall half-life of free pheromone is about 2 ms. The direct formation of FA is comparatively slow (267 ms), and FA is much more readily formed via FB (reactions 2 and 4). The formation of the activated receptor molecule FAR' via the complexes FA and FAR is relatively slow, altogether in the range of about 400 ms. Hence, it would appear that the presence of PBP acts as a preconcentrator—creating and allowing quite high concentrations of the pheromone to be delivered to the receptor.

Pheromone perception in insects is a useful model for making comparisons to vertebrate olfaction. In contrast to odor perception, pheromone perception can involve one known molecule that elicits a specific and measurable response. It is easier to measure electrophysiological and behavioral responses in vivo in insects than in vertebrates. Moreover, it is easier to identify the insect genome and produce mutant strains. Common characteristics of OBPs and PDPs suggest that they may fulfill similar functions. Nevertheless, important differences exist between the olfactory systems of these two groups of animals. Unlike mammals, the structure of the olfactory system in insects allows independent regulation of the aqueous environment (Steinbrecht 1998; Xu 2005; Kim et al. 1998). Furthermore, in contrast to the mammalian nose, which carries out a variety of functions, the antennae of insects are specialized for odor and pheromone perception (Pelosi and Maida 1990, 1995). Genome sequencing shows that mammals and nematodes express a large number of ORs and very few OBPs. Consequently, it has been proposed that in animals OBPs play a much smaller role, while the combinatorial use of ORs constitutes the main mechanism of odorant discrimination. Conversely, insects have a smaller number of ORs and a larger number of OBPs; thus, OBPs may play a more important role. In each sensillum a subset of OBPs and ORs could work together in the discrimination of odorants (Forêt and Maleszka 2006; Hekmat-Scafe et al. 2002).

1.4.2 OLFACTORY SYSTEMS AND BIOLOGICAL ALGORITHMS

The olfactory epithelium in vertebrates and the antennae in insects are the interface between the olfactory systems and the external environment. The odorant molecules bind to *receptor proteins* in *ciliary membranes* of *olfactory receptor neurons* (ORNs). An individual odor interacts with a subset of the huge number of receptors existing here, activating the ORNs linked to the subset. An ORN produces a series of electrical spikes—the frequency of which is a monotonically increasing function of the odorant concentration. This is dependent on the receptor-odorant binding affinity. As each neuron expresses one or a few receptor types (Kauer and White 2001; Korsching 2001), a pattern of active and inactive ORNs is associated with the odorant. Different odors interact with different but overlapping sets of receptors, generating different but overlapping ORN patterns. Hence, a combinatorial coding of the odor is performed by the olfactory system in this first stage of perception (Firestein 2001; Malnic et al. 1999).

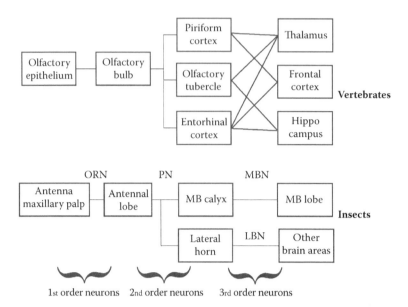

FIGURE 1.10 Parallels in olfactory processing between mammals and insects. Odorants emitted from a stimulus activate distinct subsets of ORNs in both mammals and insects (first-order neurons), and these converge on glomeruli in either the olfactory bulb or the antennal lobe (AL). From here information is relayed to higher brain centers, which have functional and neuroanatomical parallels in mammals and insects (second- and third-order neurons). The synaptic connections in the AL of insects and olfactory bulb in vertebrates are complex, and olfactory information is modified at this level. The two major target neurons of the ORNs are local interneurons (LNs), which provide lateral connections (not shown) among glomeruli and are found in both vertebrates and mammals, and cholinergic projection neurons (PNs), most of which link individual glomeruli to two higher olfactory centers, the mushroom body (MB) and the lateral horn in insects (second-order neurons). The mushroom body neurons (MBNs) project to the mushroom body lobe and other brain areas. In vertebrates the projections from the olfactory bulb go mainly to the piriform cortex, with other connections to the olfactory tubercle and entorhinal cortex and from these they connect via third-order neurons to the thalamus, frontal cortex, hippocampus, and other areas of the brain.

The odor response pattern, extended through a large number of ORNs, is then processed by the OB in mammalians or the antennal lobe in insects. These second layers are composed of spheroidal structures, *glomeruli*, made up of the synapses between ORN axons and dendrites of second-order neurons, *mitral* and *tufted* (M/T) cells in the vertebrate systems, and *projection neurons* (PNs) in the insects. M/T cells and PNs are the output channels of the olfactory bulb/antennal lobe (see Figure 1.10).

Each ORN type projects into one or two glomeruli (Wilson and Mainen 2006), and a single M/T (PN) receives input from just one type of ORN, expressing the same type of receptor, and sends its axon to the olfactory cortex. Therefore, the first representation of the odor is transduced at the glomerular level, to produce a second spatial and very ordered pattern, representing its molecular features and distributed among a much smaller number of output cells (chemotopic coding) (Mombaerts et al. 1996b; Mori et al. 1999; Raman et al. 2006a; Ressler et al. 1994; Vassar et al. 1994; Wang et al. 1998).

The convergence ratio of ORNs to M/T and PN is very high (in mammals it is around 25,000 ORNs per glomerulus), and it allows the amplification of weak signals, as well as an averaging of the receptor input. In computational terms, this averages out uncorrelated noise and leads to an increase in the signal-to-noise ratio (S/N) and, consequently, in the sensitivity of the OB/AL when compared to that of a single ORN (Duchamp-Viret et al. 1989). It is possible to detect odorants below the detection threshold of individual ORNs.

In the OB there are also interneurons, *periglomerular and granule cells*, and *local neurons* in the antennal lobe of the insects. Interneurons allow communication within and between glomeruli and regulate the activities of the M/T and PN. Periglomerular cells receive input from the ORN axons and form inhibitory synapses with M/T and PN dendrites within the same glomerulus. Granule cells form inhibitory synapses with M/T and PN of different glomeruli. Local neurons, in the antennal lobe, connect glomeruli and are mainly inhibitory (Shepherd et al. 2004). While evidence is not concrete, the interaction through PG cells may serve as a gain control mechanism, enabling the identification of odorants over several log units of concentration. Local inhibition introduces time as an additional dimension for odor coding by generating temporal patterning of the spatial code available at the glomerular (GL) layer.

The representation of odors at the glomeruli level is sent to the olfactory cortex, where different odorants elicit distinct, sparse, and distributed but partially overlapping activity patterns of the *pyramidal cells*, the main neurons of the cortex. These cells receive synapses from multiple M/T units, and their activity, consisting of action potentials, is observed only if a defined combination of second-order neurons is synchronously stimulated. Since an M/T is activated by only one type of receptor, an individual odor must excite a precise combination of receptors to generate an action in a few pyramidal neurons, implementing a coding that is sparse (Poo and Isaacson 2009). This organization leads to a more compact odorant representation than that available at the epithelium, and decouples odor quality from odor intensity.

This mechanism increases the discrimination of odors that are structurally similar. In fact, although two comparable odors generate an analogous pattern at the glomeruli level, they activate different and multiplexed pyramidal cells, generating a diverse pattern at the cortex level. Moreover, in this region, the olfactory information is combined together with the information from other sensory organs and from previous memories to give perception of the odor.

Similar information processing mechanisms are observed in the mushroom body of insects. The projection neurons in the antennal lobe synapse with Kenyon cells (KCs), the main neurons in the mushroom body. In the locust 50,000 KCs have been identified. Each of these receives input from more PNs, and each PN synapses with about 600 KCs. In this way, the chemotopic representation of an odor, obtained at the glomerulus level by the broadly tuned PNs, is transduced into a sparse code generated by the highly specific KCs (Szyszka et al. 2005). Schematic representations of the excitatory responses of distinct second- and third-order neurons to

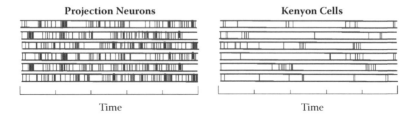

FIGURE 1.11 Excitatory responses of second- and third-order neurons. The second-order neurons, M/T units in mammalians and PNs in insects, generate a second representation of odors at the glomeruli level and send it to the olfactory cortex or mushroom body, where third-order neurons are stimulated (Kenyon cells in insects). The activity of these third-order cells is observed only if a defined combination of the second-order neurons is synchronously stimulated. Since an M/T or PN is activated by one type of receptor only, an individual odor must excite a precise combination of receptors to generate an action in the third-order neurons in the cortex or mushroom body. This mechanism generates a sparse code of the stimulus and increases the discrimination of odors that are structurally similar. In fact, although two similar odors may generate an analogous pattern at the glomeruli level, they activate different cells in the cortex or mushroom body (decorrelation). The figure shows a representation of firing patterns of different cells at the projection neuron level on the left, and the much lower level of firing observed at the third-order neuron level on the right.

different odors are shown in Figure 1.11. Perez-Orive et al. (2002) observed that even though a sparse coding mechanism makes the system more vulnerable to damage, it introduces important benefits. With respect to a nonsparse coding mechanism, it increases the discrimination power, enhances the olfaction system's memory capacity, and makes easier the association between patterns generated by using a few neurons and memories previously stored.

1.5 ARTIFICIAL OLFACTION

There has been a large growth of interest in building systems that imitate the five senses of the mammal. The sense of smell is perhaps the least appreciated in man's daily life, while many animals rely heavily on their acute sense of smell. Nevertheless, it is an invaluable tool in many industries, past and present. An artificial sensing system that mimics the human sense of smell is desired in a range of fields, such as food and drink production, tobacco and cosmetics industries, and for environmental monitoring in chemical plants. The human sense of smell is influenced by many factors, such as age, gender, state of health, and mood. The measure of sense is limited to linguistic ability of the individual in communication between one person and another. The response of subjects to different odors and odor concentrations is highly subjective, as the human sense of smell is relatively weak and can vary dramatically from person to person. An instrument that could perform simple odor discrimination and provide measurement of odor intensity, without the influences mentioned above, would be very useful in modern industry. The concept of an electronic nose has systematic similarities to the way humans sense odors. Sensors in the system act as receptors in the nasal mucosa, and the responses given by the sensors form a pattern (a set of signals), like the resulting event in the olfactory bulb due to a stimulus. As outlined in the previous sections, the biological olfactory code is not yet well understood, but the basic criteria of a nose for odor recognition are broadly defined. This information was sufficient for designing a machine with which it was possible to discriminate and recognize odors. One of the first instruments specifically designed to detect odors was developed in 1961 (Moncrieff 1961), which was really a mechanical nose. An electronic nose was reported in 1964 by Wilkens and Hartman (1964) based on redox reactions of odorants at an electrode. In the following year, Buck et al. (1965) and Dravnieks and Trotter (1965) showed the potential of electronic noses based on the modulation of conductivity and the modulation of contact potential by odorants, respectively. Research in this field showed little progress until 1982 when Persaud and Dodd (1982) reported a successful discrimination of a wide variety of odors using plural semiconductor transducers. The concept of an electronic nose, as an intelligent chemical array sensor system, for odor classification was practically modeled for the first time. The heart of any electronic nose is the type of sensors used for the particular system. Applications of the electronic nose are typically restricted by the sensors' characteristics to the target gases or odors and the operating environment. Many types of sensor have been developed to detect specific gases and vapors since the 1970s. The human nose can identify many odors that may contain hundreds of individual chemical components, and therefore the sensors for an electronic nose should be generalized at the molecular level. The desired properties for sensors are high sensitivity, rapid response, good reproducibility, and reversibility to large numbers of chemicals. It is also better for an electronic nose to be small in size, flexible, and able to adapt to many environments as well as operate at ambient temperatures.

There are numerous types of gas sensor arrays used in electronic noses. These include metal oxide semiconductor (MOS) sensors, catalytic gas sensors, solid electrolyte gas sensors, conducting polymers, mass-sensitive devices, and fiber-optic devices based on Langmuir–Blodgett films (see Table 1.1). The oxide materials, which have been popular for use in electronic noses, operate on the basis of modulation of conductivity when the odorant molecules react with chemisorbed oxygen

TABLE 1.1
Sensor Devices: Example Technologies Used in Gas Sensor Arrays

Sensor Type	Transduction Principle
ChemFET, light addressable potentiometric sensors	Work function
Chemoresistors	Conductivity
Amperometric gas sensors	Ionic current
Chemocapacitors	Permittivity
Thermopile, pellistor catalytic sensor	Temperature
Colorimeter, spectrophotometer	Optical spectrum
Optical fibers	Fluorescence, absorption
Optical fibers, surface plasmon resonance	Refractive index
Cantilevers, surface acoustic wave, quartz crystal microbalance	Mass

species. There are commercially available metal oxide sensors that operate at elevated temperatures, between 100 and 600°C (Arshak et al. 2012). They are sensitive to combustible materials (0.1–100 ppm), such as alcohols, but are generally poor at detecting sulfur- or nitrogen-based odors and have a major problem of irreversible contamination with these compounds. Integrated thin-film metal oxide sensors have been designed using planar integrated microelectronic technology that has advantages of lower power consumption, reduction in size, and improved reproducibility; however, they tend to suffer from poor stability. There are a number of advantages in employing organic materials in electronic noses. A wide variety of materials are available for such devices, and they operate close to or at room temperature (20–60°C) with a typical sensitivity of around 0.1–100 ppm. Furthermore, functional groups that interact with different classes of odorant molecules can be built into the active material, and the processing of organic materials is easier than that of oxides.

Persaud and Pelosi (1985) proposed an electronic nose using conducting polymers after investigating properties of a number of conducting polymers. They have found several organic conducting polymers that respond to gases with a reversible reaction of conductivity, fast recovery, and high selectivity toward different compounds. In an experiment with an array of 5 different conducting polymer sensors and 28 odorants, they observed 20 different sets of responses and showed possible discrimination with 14 of the odorants by measuring changes in the electrical resistance. These results led Persaud (2005) to produce arrays of gas-sensitive polymers that had reversible changes in conductivity and rapid adsorption/desorption kinetics at ambient temperatures when they were exposed to volatile chemicals. The concentration-response profiles of such sensors are almost linear over a wide concentration range to single chemicals. This is advantageous, as simple computational methods may be used for information processing. The raw signals from a sensor array consisting of 32 conducting polymers are shown in Figure 1.12. It can be seen that all the sensors respond to the pulse of the odorant, but there are a range of responses—each sensor having a different degree of response to the same concentration of odorant. This is dependent on the selectivity and sensitivity of each element of the sensor array to the analyte.

1.5.1 DATA PROCESSING

A set of output signals from individual sensors within an electronic nose must be converted into signals appropriate for computer processing. Electronic circuits within a sensing instrument usually perform an analog-to-digital conversion of output signals from sensors, feature extraction of useful information, and interfacing to an external computer for pattern analysis. For unattended automated sampling, it is common to have a mass flow control system that delivers the odor from the source to the sensor array. The final response vectors generated by the sensors are then analyzed using various pattern recognition techniques.

Early researchers in the field used various multisensor systems for identification of different types of gases, often in parallel with analytical methods. Zaromb and Stetter (1984) provided a theoretical basis for the selection and effective use of an array of chemical sensors for a particular application. Bott and Jones (1986) attempted to build a multisensor system to monitor hazardous gases in a mine using six sensors of three

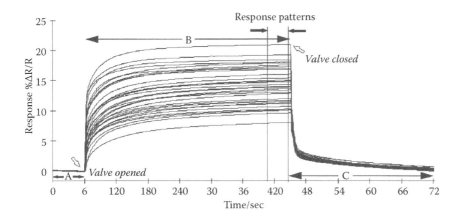

FIGURE 1.12 Raw data from an array of conducting polymer sensors. The raw data responses over time to an odorant stimulus (in this case acetone) from an array of 32 conducting polymer sensors are shown. The array is exposed to clean air and a baseline is taken (A). On opening a valve to present the stimulus, a change in resistance in each sensor is observed (B), shown as %ΔR/R, where R is the base resistance in clean air and ΔR the change observed. This response is dependent on the sensitivity and selectivity of each individual sensor to the odorant. When the valve is closed and clean air is presented again to the sensor array, the resistances gradually return to baseline (C) as molecules desorb from the sensor surfaces. In the region between cursors we can extract normalized response patterns that can be used for further data processing (see Figure 1.14).

different types in combination with oxidizing layers and absorbent traps. The system was able to distinguish between gases evolved from a fire and those evolved from diesel engines or explosives. Müller and Lange (1986) demonstrated possible identification and concentration measurement of six different gases by means of four MOS gas sensors with layers of different types of zeolite filters. Gall and Müller (1989) adapted similar methods for identifying gas mixtures using partial least-squares and transformed least-squares methods for analysis, which led Sundgren and coworkers (1990) to attempt improvements using three pairs of Pd-gate MOSFETs and Pt-gate MOSFETs. Kaneyasu et al. (1987) modeled an early version of an electronic nose using six integrated sensor elements with a single chip microcomputer. Odor-identifying systems containing two or more different types of sensors are often an attempt to enhance the different dimensional characteristics of the responses. Stetter and coworkers (1986) demonstrated a combined system with a hydrocarbon sensor and an electrochemical sensor (Stetter et al. 1990). They managed to get responses from both the hydrocarbon sensor and the electrochemical sensors after passing the vapors through a combustible gas sensor. The resulting data were successfully analyzed using pattern recognition methods based on neural networks.

Despite a number of disadvantages, including high power consumption, elevated operational temperature, poisoning effects from sulfur-containing compounds, and poor long-term stability, MOS gas sensors are the most widely used in gas and odor detection. The main reason is that MOS commercial products have been available for a number of years. Abe et al. examined an automated odor sensing system based on plural semiconductor sensors to measure 30 substances (Abe et al. 1987) and 47 compounds (Abe et al. 1988). They analyzed the sensor outputs using pattern recognition techniques: Karhunen–Loeve (K-L) projection for visual display output and k-nearest neighborhood (k-NN) method and potential function method for classification. Shurmer et al. (1989) worked on discrimination of alcohols and tobaccos using tin oxide sensors based on the correlation coefficient method in their research. Weimar et al. (1990) demonstrated the possibility of determining single gas components, such as H_2, CH_4, and CO, in air from specific patterns of chemically modified tin oxide-based sensors by using two different multicomponent analysis approaches. Most methods applied to identification, classification, and prediction of gas sensor outputs were based on conventional pattern recognition techniques until the late 1980s. In the 1990s Sleight (1990) and Gardner et al. (1990) suggested the possible application of artificial neural networks (ANNs) to electronic nose systems. Gardner and coworkers implemented a three-layer back-propagation network with 12 inputs and 5 outputs architecture for the discrimination of several alcohols, where they reported that it was better than the previous work (Shurmer et al. 1990) carried out using analysis of variance (ANOVA). Cluster analysis and principal component analysis (PCA) were used to test 5 alcohols and 6 beverages from 12 tin oxide sensors (Gardner 1991; Gardner et al. 1992a). The results were presented by raw and normalized responses, and showed that the theoretically derived data normalization substantially improved the classification of chemical vapors and beverages. Further investigations were carried out to discriminate the blend and roasting levels of coffees, the differences between tobacco blends in cigarettes, and three different types of beer. The result confirmed the potential application in an electronic instrument for on-line quantitative process control in the food industry (Shurmer and Gardner 1992; Gardner et al. 1992a, 1992b). Hines and Gardner (1994) also developed a standalone microprocessor-based instrument that can classify the signals from an array of odor-sensitive sensors. Data from the odor sensor array were initially trained on a personal computer using a neural network program, and then the neuronal weights were sent to an artificial neural emulator (ANE), which consisted of a microprocessor, ADC chips, read only memory (ROM), and random access memory (RAM). The group also attempted to improve the performance of oxide semiconductor sensors, with respect to long-term stability and poisoning effect (Gardner 1995), a multisensor system using an array of conducting polymers (Gardner et al. 1994), and an adaptation of fuzzy neural networks for classification (Singh et al. 1996).

Another approach to odor sensing was studied using a quartz resonator sensor array where the mechanism of odor detection is based on the changes in oscillation frequencies when gas molecules are adsorbed onto sensing membranes. Nakamoto et al. employed neural network pattern recognition, including three-layer back-propagation and principal component analysis, for the discrimination of several different types of alcoholic drinks using a selection of sensing membranes (Ema et al. 1989; Nakamoto et al. 1990, 1991). They also proposed a new processing element model based on fuzzy theory and Kohonen's learning vector quantization (LVQ) for the discrimination of known and unknown odors (Moriizumi et al. 1991).

Pattern recognition in the electronic nose system may be regarded as a branch of artificial intelligence that involves the mimicking of human intelligence to solve

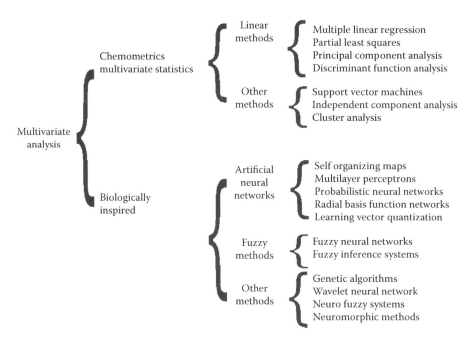

FIGURE 1.13 Data processing methods. The diagram illustrates a number of methodologies used in processing multidimensional data from sensor arrays.

chemical problems. Figure 1.13 summarizes methods commonly used in the field, some of which are discussed in more detail here. There are two main approaches to pattern recognition: parametric and nonparametric. Parametric methods rely upon obtaining or estimating the probability density function of the parameters used to characterize the response of a system. Conversely, nonparametric methods require no assumption about the fundamental statistical distributions of data. Two types of nonparametric learning or classification methods are available: supervised and nonsupervised. Supervised methods involve the learning of data based on advance knowledge of the classification, whereas unsupervised methods make no prior assumption about the sample classes but try to separate groups or clusters. Mapping methods in pattern recognition are an unsupervised way of analyzing chemical inputs. Many authors since these early days of electronic nose research have used pattern recognition techniques for a number of practical applications. These include wine discrimination (Garcia et al. 2006; Lozano et al. 2008), identification of microorganisms (Moens et al. 2006), fish freshness assessment (GholamHosseini et al. 2007), urinary tract detection (Kodogiannis et al. 2008), and identification of components in mixtures (Penza and Cassano 2003; Penza et al. 2002).

1.5.2 ALGORITHMS

Data projection or mapping is a highly desirable feature in gas and odor sensing, as it allows an approximate visual examination of multidimensional input data. The patterns from the electronic nose system due to chemical stimuli are described in a multidimensional space (this depends on the number of features used). Human vision is very good at recognizing patterns in small dimensions (≤3), but cannot perceive high-dimensional (≥3) relationships. Data projection enables us to visualize high-dimensional data to better understand the underlying structure, explore the intrinsic dimensionality, and analyze the clustering tendency of multivariate data. There are a number of ways to achieve this, but the basic object is the same, which is to reduce the original high dimension to a lower dimension, preserving the structure of the input patterns as well as possible.

Two general approaches are available for the mapping methods: linear and nonlinear. In linear mapping, each of the new features described in a low dimension (usually either two- or three-dimensional space) is a linear combination of the original features that are described in a higher dimension. A linear mapping function is well defined, and its mathematical properties are well understood. Linear projection methods have been commonly used in the pattern recognition parts of many gas and odor sensing systems. Stetter et al. (1986) applied K-L transformation (reviewed by Wold et al. 1987) to project 20-dimensional hazardous gases and vapors onto two-dimensional space for the analysis. Gardner and coworker (Gardner 1991; Shurmer and Gardner 1992) applied PCA for the discrimination of chemical data from their multisensor array and a hybrid electronic nose. Abe et al. (1987, 1988) used K-L projection for the visualization of the data collected with their plural gas sensors and attempted to classify by applying extra pattern recognition techniques such as cluster analysis. Nakamoto et al. (1991) used the first and second principal components in a two-dimensional diagram to show the fine differences among whisky aromas.

Simplicity and generality are the main advantages of linear mappings, whereas nonlinear mapping algorithms have complicated mathematical formulations. Nonlinear mappings have been rarely applied in gas and odor sensing; however, they can be used when the preservation of complex data structures has failed with linear mappings. Kowalski and Bender (1972, 1973) applied nonlinear mappings as well as linear mappings to display chemical data in either two or three dimensions. They have also studied various mapping methods and suggested a combined linear and nonlinear mapping method. Self-organizing maps (SOMs), based on the Kohonen network, have also been shown to be very useful for discrimination of odor classes (Bona et al. 2012).

If we look at the raw data shown in Figure 1.12, we can extract features from it. By normalizing the raw data so as to express each individual sensor response as a fraction of the total response of the entire array of sensors, we can produce a series of patterns that represent individual classes of analytes presented to the array. Such a series of patterns are shown in Figure 1.14. These patterns represent vectors that may be used as the inputs for further processing of data for classification.

1.5.2.1 Sammon Mapping

The nonlinear mapping method proposed by Sammon (1969, 1970) is based upon a point mapping of N L-dimensional vectors from the L-space to a lower-dimensional space such that the inherent data structure is approximately preserved. Therefore, the resultant data configuration can be easily evaluated by human observations in either two or three dimensions.

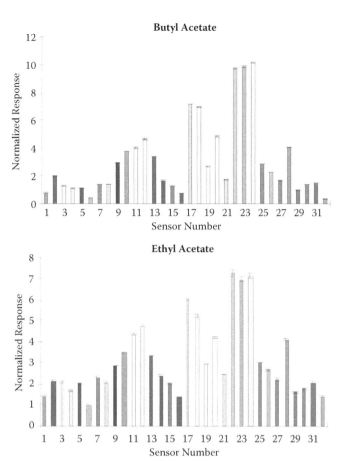

FIGURE 1.14 Patterns generated from an array of sensors. The raw responses (Figure 1.12) to different analytes can be normalized to be relatively independent of concentration. As a result of the differences in sensitivity and selectivity of individual sensors in the array to different analytes, patterns of responses can be seen that allow visual discrimination between analytes. These data vectors form the input data to further data analysis and pattern recognition. Normalized response patterns of single volatile chemicals are shown. Error bars represent standard deviations from five repeated measurements.

For N number of patterns in L-dimensions designated X_i ($i = 1, 2, \ldots, N$) and corresponding N patterns in d-dimensions ($d = 2$ or 3) designated Y_i ($i = 1, 2, \ldots, N$), let the distance measure between the patterns X_i and X_j in L-dimensions be defined by $d_{ij}{}^* = dist[X_i, Y_j]$ and the distance measure between the corresponding patterns Y_i and Y_j in d-dimensions be defined by $d_{ij} = dist[X_i, Y_j]$. Any distance measure functions could be used for the interpattern distances, such as the generalized Mahalanobis distance,

$$D_{ij} = \left[\sum_{k=1}^{m} \left(Y_{ik} - Y_{jk} \right)^{P} \right]^{1/P}$$

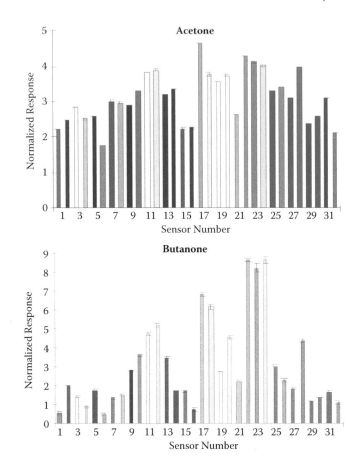

FIGURE 1.14 (continued) Patterns generated from an array of sensors. The raw responses (Figure 1.12) to different analytes can be normalized to be relatively independent of concentration. As a result of the differences in sensitivity and selectivity of individual sensors in the array to different analytes, patterns of responses can be seen that allow visual discrimination between analytes. These data vectors form the input data to further data analysis and pattern recognition. Normalized response patterns of single volatile chemicals are shown. Error bars represent standard deviations from five repeated measurements.

Hamming metric,

$$H_{ij} = \sum_{k=1}^{m} \left| Y_{ik} - Y_{jk} \right|$$

and the most common Euclidean distance,

$$d_{ij}^* = \left[\left(X_{ik} - X_{jk} \right)^2 \right]^{1/2}$$

The Y patterns have the following configuration, where the initial values of Y patterns in d-space are generated with random values: $Y_1 = [y_{11}, y_{12}, \ldots, y_{1d}]^T$, $Y_2 = [y_{21}, y_{22}, \ldots, y_{2d}]^T$, and $Y_N = [y_{N1}, y_{N2}, \ldots, y_{Nd}]^T$, where T denotes transpose of the vector.

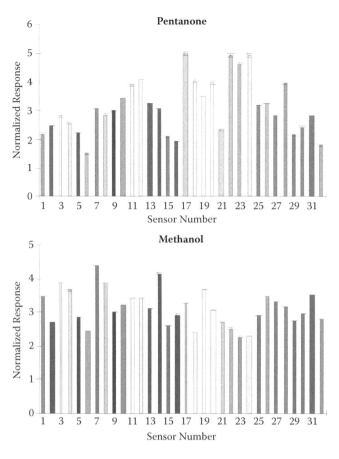

FIGURE 1.14 (continued) Patterns generated from an array of sensors. The raw responses (Figure 1.12) to different analytes can be normalized to be relatively independent of concentration. As a result of the differences in sensitivity and selectivity of individual sensors in the array to different analytes, patterns of responses can be seen that allow visual discrimination between analytes. These data vectors form the input data to further data analysis and pattern recognition. Normalized response patterns of single volatile chemicals are shown. Error bars represent standard deviations from five repeated measurements.

All the d-space interpoint distances d_{ij} are then used to define an error E, which represents how well the present configuration of N points in the d-space fits the N points in the L-space, i.e.,

$$E = \frac{1}{\sum_{i<j}\left[d_{ij}^*\right]} \sum_{i<j}^{N} \frac{\left[d_{ij}^* - d_{ij}\right]^2}{d_{ij}^*} \tag{1.3}$$

The error term is a function of the $d \times N$ variables y_{pq} ($p = 1, 2, ..., N$) and ($q = 1, 2, ..., d$), where the y_{pq} variables are adjusted or, equivalently, the d-space configuration is changed so as to decrease the error. A steepest descent procedure, which is a gradient method, is used to search for a minimum of error E. It is clear from

Equation 1.3 that the error is only zero when $d_{ij}^{*} = d_{ij}$, which can only be accomplished when the intrinsic dimensionality is actually two- or less dimensional (three or less for a three-dimensional mapping). In general, the nonlinear mapping attempts to emphasize the fit to small L-space distances since the d_{ij}^{*} term in the denominator of the error function acts as a weighting function. which is large when d_{ij}^{*} is small. Once the error is minimized to an acceptable level, the resultant patterns may then be plotted on a two- or three-dimensional graph for the visualization.

1.5.2.2   Combination of Sammon's Nonlinear Mapping with PCA Linear Projection Method

The original Sammon's nonlinear mapping algorithm initializes the configuration of Y patterns with small random values, where the configuration of Y patterns is updated closer to the optimum position each time an iteration is completed. It is clear that different initial configurations will result in different output configurations of patterns in the projection space, as Sammon's nonlinear mapping is based on the preservation of interpoint distances of patterns. In addition, a rotation of the resulting map may be observed when new patterns are introduced into the data set. These create difficulties for the observer to relate one map to another. An attempt was made to tackle the listed problems by combining principal component analysis (PCA) linear projection with Sammon's nonlinear mapping method, which also provided a more validated initialization in the practical application sense.

PCA is popular and a very widely used linear projection method applied in pattern recognition and data analysis. It is based on the statistical Karhunen–Loeve transformation, which creates new variables as linear combinations of the original variables. A linear mapping can be conveniently expressed as $Y_i = A \bullet X_i$; $i = 1, 2, ..., N$; where Y_i is a $d \times 1$ vector, X_i is a $L \times 1$ vector, and A is a $d \times L$ matrix. This involves a linear orthogonal transform from a L-dimensional input space to a d-dimensional space, $d \leq L$, such that the coordinates of the data in the new d-dimensional space are uncorrelated and a maximal amount of variance of the original data is preserved by only a small number of coordinates. The rows of the matrix A are the eigenvectors corresponding to the d largest eigenvalues of the $L \times L$ covariance matrix C; so, in the case of a two-dimensional projection, the eigenvectors corresponding to the two largest eigenvalues are the rows of matrix A.

The output patterns from the PCA algorithm are then scaled down to smaller values to be used as initial values of Sammon's nonlinear mapping. However, the output patterns, from the above steps, could be implemented alone for a linear projection method, as they are linear combinations of all the original coordinates, and this is often the case found in papers published in the electronic nose field. Since each eigenvalue is proportional to the variance along its corresponding eigenvector, a measure is available of the percent variance retained by the chosen principal components. Equation 1.4 shows percent variance in the two-dimensional space case:

$$\%V = \frac{(\lambda_1 + \lambda_2)}{\sum_{i=1}^{L} \lambda_i} \times 100 \tag{1.4}$$

The value indicates how well the interstructure is preserved by the principal components from the original multidimensional space. Once the NL-dimensional space patterns are related to the patterns in d-dimensional space ($d = 2$ or 3), the resultant patterns are presented on a scatter graph for observation.

It has been shown that relatively simple implementations of biologically justified neural rules for Hebbian learning can produce PCA; hence, there has been much interest in these applications (Karhunen and Joutsensalo 1995). Linear approaches have been shown to have several limitations, and introduction of nonlinearities in mapping between input and output data is advantageous.

Figure 1.15 shows a principal component analysis of patterns taken from Figure 1.14. It can be seen that discrete clusters appear representing individual analytes presented to the array. If this is used to initialize a Sammon map, the resultant graph is shown in Figure 1.16, where now the axes are associated with a Euclidean distance measure between each pattern input into the map.

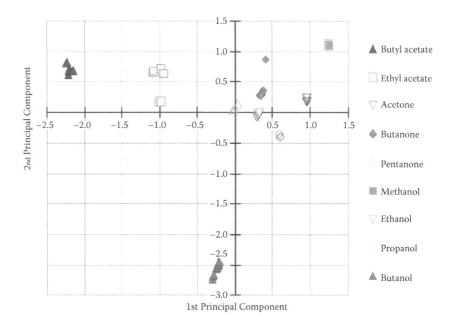

FIGURE 1.15 Principal component analysis of patterns. Data projection or mapping is a highly desirable feature in gas and odor sensing, as it allows an approximate visual examination of multidimensional input data. The patterns from the sensor array generated as a result of chemical stimuli are described in a multidimensional space (depending on the number of features used). Human vision is very good at recognizing patterns in small dimensions (<3), but cannot perceive high-dimensional (>3) relationships. Data projection enables us to visualize high-dimensional data to better understand the underlying structure. Raw data recorded from a sensor array were normalized and principal component analysis was carried out, and the graph plots principal component 1 versus principal component 2. Each point on the graph corresponds to one measurement. It is seen that different analytes cluster together in different locations on the graph. Here are shown butyl acetate, ethyl acetate, acetone, butanone, pentanone, methanol, ethanol, propanol, and butanol, all measured using a 32-sensor array.

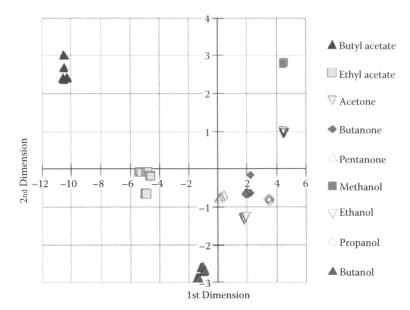

FIGURE 1.16 Sammon map. The Sammon map allows visualization of the relationship between one pattern and another based on a distance measure such as the Euclidean distance. It may be more informative than a PCA where the axes do not necessarily indicate if different patterns are separable from each other. The original Sammon's nonlinear mapping algorithm initializes the configuration of Y patterns with small random values, where the configuration of Y patterns is updated closer to the optimum position each time an iteration is completed. The PCA can be used to initialize a Sammon map, and this allows rapid convergence as well as invariance in the representation of the relative distance between patterns.

1.5.2.3   Competitive Learning Neural Networks

Classification of input chemicals is one of the main functions that pattern recognition methods provide in an electronic nose system. The visualization of sampled chemicals gives information on the interrelationship between input patterns via human observation. However, it is important in an intelligent system that the ability to classify does not rely on human judgment, so producing a system that can be truly automated.

There are many ways to achieve classification using pattern recognition methods. However, parametric methods of pattern recognition (classical statistical pattern recognition methods), which assume the probability density function is known or can be estimated, do not consider the many problems associated with practical applications and are rarely used in chemical analysis. In most cases there are two stages used in a pattern recognition process. First, the response patterns from an electronic nose are trained by a pattern recognition method using mathematical rules that relate the patterns to a known class (training or learning stage). Then the response from an unknown chemical pattern is tested against the knowledge base and the classification is given (testing or recognition stage). This kind of process is based on supervised learning, which is a nonparametric pattern recognition method. Some of the supervised learning methods are linear techniques and assume that the response vectors are well described in Euclidean space. Distance functions are often used to describe a measure of similarity between patterns through their proximity. However, this kind of measure is only useful when the sensor outputs are linearized or concentration independent.

Artificial neural networks (ANNs) are more sophisticated methods of solving chemical problems, which had not been applied to gas and odor sensing using multi-sensor arrays until late in the 1980s (Gardner et al. 1990, 1992a; Sleight 1990). The ANNs were developed to provide models that were able to represent some aspects of the working principles of the brain, in particular, learning from experience. These methods can handle nonlinear data and offer potential advantages, such as fault tolerance to sensor drift or noise, adaptability, and high data processing rates, over classical pattern recognition methods (Gardner et al. 1990, 1992a; Persaud and Pelosi 1992). A neural network is characterized by three basic elements: the process units that represent the neurons, the network architecture (neuron connections) where the inputs are individually weighted and the output determined by an activation function, and a learning rule that provides the law according to which the network can learn from experience. Figure 1.17 shows a typical neuron model with synaptic connections and a simple processing unit that is capable of performing nonlinear transformations. This type of network is called the generalized perceptron. The neuron in the network is a processor unit with multiple inputs and one output, where the processor unit is divided into two parts: the first part is a weighted sum of the inputs, and the second is a nonlinear transformation of the sum. The weights are optimized according to the learning rule as a function of the network's experience.

A multilayer perceptron network is probably the most widely used architecture, as it can be applied to many problems (Gardner et al. 1992a; Kodogiannis et al. 2008). Figure 1.18 shows a three-layer feed-forward network where the neurons are arranged in three layers: input, hidden, and output layers. The neurons in a layer

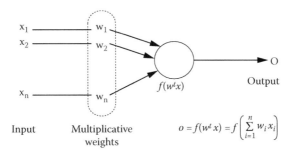

$$o = f(w^t x) = f\left(\sum_{i=1}^{n} w_i x_i\right)$$

FIGURE 1.17 Perceptron model. A typical neuron with synaptic connections and a simple processing unit that is capable of performing nonlinear transformation is shown. This type of network is called the generalized perceptron. The neuron in the network is a processor unit with multiple inputs and one output, where the processor unit is divided into two parts: the first part is a weighted sum of the inputs, and the second is a nonlinear transformation of the sum. The weights are optimized according to the learning rule as a function of the network's experience. An input vector $x_1 \dots x_n$ is multiplied by a weight vector $w_1 \dots w_n$ and summed to produce a weighted output. This forms a single processing element.

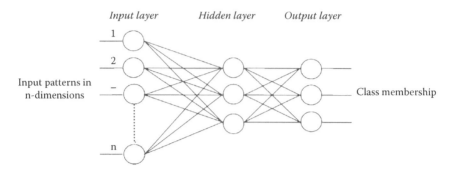

FIGURE 1.18 Multilayer perceptron network. Each perceptron may be interconnected with other perceptrons to produce a network of processing elements. Here is a three-layer feed-forward network where the neurons are arranged in three layers: input, hidden, and output layers. The neurons in a layer are connected to all the neurons in the following layers, and the weights exist in every connection between two layers. Once the network topology is defined, a set of input–output data is used to train the network to determine appropriate values for the weights associated with each interconnection. The data are propagated forward through the network to produce an output that is compared with the corresponding output (target) in the data set to obtain an error. This error is minimized by updating the weights through an iterative process, until the network has reached an optimal convergence.

are connected to all the neurons in the following layers, and the weights exist in every connection between two layers. The number of neurons in the input and output layers depends on the respective number of inputs (dimensions of input patterns) and outputs (number of classes) being considered; however, the number of neurons in the hidden layer is chosen by the user, which essentially defines the topology of the network. Each interconnection has an associated weight that modifies the strength of the signal flowing along that path. Thus, with the exception of the neurons in the input layer, the input to each neuron is a weighted sum of the outputs from neurons in the previous layer. The output of each node is obtained by passing the weighted sum through a nonlinear operator, which is typically a sigmoidal function.

Once the network topology is defined, a set of input-output data is used to train the network to determine appropriate values for the weights associated with each interconnection. The data are propagated forward through the network to produce an output that is compared with the corresponding output (target) in the data set to obtain an error. This error is minimized by updating the weights through an iterative process, until the network has reached an optimal convergence. The last set of weights is retained as the parameters of the neural network model. Process modeling using ANNs is very similar to identifying the coefficients of a parametric model of specified order, where the magnitudes of the weights define the characteristics of the network. However, unlike conventional parametric model forms, which have an a priori assigned structure, the weights of an ANN also define the structural properties of the model. Thus, an ANN has the capability to represent complex systems whose structural properties are unknown. The back-propagation learning method by Rumelhart et al. (1986) has been used widely in various applications, as it was one of the first effective learning techniques introduced for the neural network models.

Typically for the back-propagation training algorithm, the mean square error method is used as a measure of error during the iterative learning process, where the gradient descent (steepest descent) method tries to minimize the network total error by adjusting the weights. The negative gradient of the error function, with respect to the weights, then points in the direction which will most quickly reduce the error function, and so the final minimum is reached when the gradient becomes zero.

In contrast to supervised learning, pattern recognition methods with unsupervised learning do not require a separate training stage. They learn to discriminate between the response vectors automatically by making no prior assumption about the sample classes but trying to separate groups or clusters. Self-organizing artificial neural networks or self-organizing feature maps (SOFMs) (Kohonen 1982) are unsupervised pattern recognition methods whose learning methods are based on a competitive learning mechanism called winner takes all. A SOM is a network of neurons arranged as knots of a planar square lattice where each neuron has four logic immediate neighbors. Each neuron is located by a vector whose components are the knot coordinates in the lattice. It has multiple input channels and one output channel whose value can be either 1 (active) or 0 (inactive). The SOM technique tries to transform input patterns of multidimensions into one- or two-dimensional discrete maps, and also to perform this transformation adaptively in a topological ordered fashion.

SOM networks are usually applied to show the input pattern distribution maps and their scatter plots. A probability density function is normally adapted for the classification of input patterns. Di'Natale and coworkers (1995a, 1995b) have applied the SOM technique for the discrimination of chemical patterns, where they have proposed a new classification method that generalizes the potential function method to neural implementation in an unsupervised environment. Distante and coworkers (Distante et al. 2000, 2002; Zuppa et al. 2004) have taken this concept further to introduce multiple self-organizing maps (mSOMs) as a powerful method for classification and feature extraction. This concept of mSOM has been used to counteract the effects of drift over time, which is a common feature of chemical sensing systems.

1.5.2.4 Self-Organizing Feature Maps

In a competitive learning neural network, the output neurons compete among themselves to be activated or fired, with the result that only one output neuron is on at any one time. Figure 1.19 shows the architecture of Kohonen's SOFM network in one dimension, and Figure 1.20 in two dimensions (B), consisting of two layers—an input layer and an output layer. Each input layer neuron has a feed-forward connection to each output layer neuron (note that only one input neuron connection is shown in Figure 1.19). The output neurons that win the competition are called winner-takes-all neurons, where a winner-takes-all neuron is chosen by selecting a neuron whose weight vector has a minimum Euclidean distance (or maximum similarity) from the input vector.

The Kohonen network performs clustering through competitive learning. The node with the largest activation level is declared the winner in the competition. This node is the only node that will generate an output signal, and all other nodes are suppressed to zero activation level. Furthermore, this node and its neighbors are the

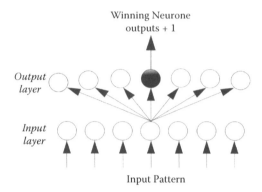

FIGURE 1.19 Kohonen SOFM networks. Self-organizing artificial neural networks or self-organizing feature maps (SOFMs) are unsupervised pattern recognition methods whose learning methods are based on a competitive learning mechanism called *winner takes all*. An SOFM is a network of neurons arranged as knots of a planar square lattice where each neuron has four logic immediate neighbors. Each neuron is located by a vector whose components are the knot coordinates in the lattice. It has multiple input channels and one output channel whose value can be either 1 (active) or 0 (inactive). The SOM technique tries to transform input patterns of multidimensions into one- or two-dimensional discrete maps, and also to perform this transformation adaptively in a topological ordered fashion. The figure shows the one-dimensional case.

only nodes permitted to learn for the current input pattern. The Kohonen network uses intralayer connections (see Figure 1.21a) (note that the latter connections are shown only for the neuron at the center of the array) to moderate this competition. The output of each node acts as an inhibitory input to the other nodes but is actually excitatory in its neighborhood. Thus, even though there is only one winner node, more than one node is allowed to change its weights. This scheme for moderating competition within a layer is known as lateral feedback. The inhibitory effect of a node can also decrease with the distance from it and assumes the appearance of a Mexican hat, as seen in Figure 1.21b. The size of the neighborhood varies as learning continues, where it starts large and is gradually reduced, making the range of change sharper and sharper.

In an SOFM, the neurons are placed at the nodes of a lattice that is usually one- or two-dimensional, although higher-dimensional maps are also possible. The neurons become selectively tuned to various input patterns or classes of input patterns in the course of a competitive learning process. The location of the neurons so tuned, i.e., the winning neurons, tend to become ordered with respect to each other in such a way that a meaningful coordinate system for different input features is created over the lattice. An SOFM is therefore characterized by the formation of a topographic map of the input patterns, in which the spatial locations of the neurons in the lattice correspond to intrinsic features of the input patterns. This map has similarity with the information processing infrastructure of the nervous system. The self-organization model is effective for dealing with complex problems whose mathematical forms are

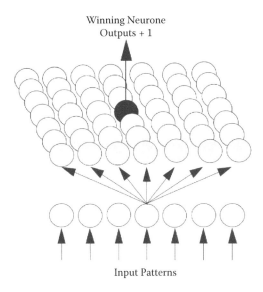

Winning Neurone
Outputs + 1

Input Patterns

FIGURE 1.20   Kohonen networks—intralayer connections. In a competitive learning neural network, the output neurons compete among themselves to be activated or fired, with the result that only one output neuron is on at any one time. Figure 1.19 shows the architecture of Kohonen's SOFM network in one dimension. Here a two-dimensional network is shown, which consists of two layers: input layer and output layer. Each input layer neuron has a feed-forward connection to each output layer neuron (note that only one input neuron connection is shown). The output neurons that win the competition are called winner-takes-all neurons, where a winner-takes-all neuron is chosen by selecting a neuron whose weight vector has a minimum Euclidean distance (or maximum similarity) from the input vector.

too complicated to define. In practice, the self-organization model compensates for inaccuracies and noise in the sensors.

In pattern classification, the requirement is to classify the input data sets into a finite number of classes such that the average probability of misclassification is minimized. This is an important task that needs to delineate the class boundaries where decisions are made. A parametric method is commonly adapted for classical pattern classification, which typically assumes a Gaussian distribution, whereas in a non-parametric method, such as SOFM, it is achieved by exploiting the density-matching property of the map. However, it is emphasized that the feature map is intended only to visualize metric-topological relationships of input patterns, and it was not recommended to use the SOFM for classification itself, as the recognition accuracy could be significantly increased if the map is fine-tuned with a supervised learning scheme (Kangas et al. 1990).

1.5.2.5   Self-Organizing Map Algorithm

The SOM may be categorized as a nonlinear projection of the probability density function of n-dimensional input data into a one- or two-dimensional lattice of output layer neurons, which comprises the output space such that a meaningful topological

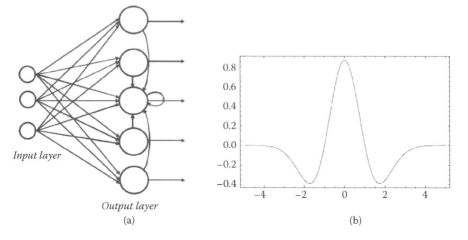

FIGURE 1.21   (a) Lateral feedback in a one-dimensional lattice Kohonen layer. (b) Mexican hat function showing inhibition and excitation areas. The Kohonen network uses intralayer connections (a) to moderate this competition (note that the latter connections are shown only for the neuron at the center of the array). The output of each node acts as an inhibitory input to the other nodes but is actually excitatory in its neighborhood. Thus, even though there is only one winner node, more than one node is allowed to change its weights. This scheme for moderating competition within a layer is known as lateral feedback. The inhibitory effect of a node can also decrease with the distance from it and assumes the appearance of a Mexican hat (b). The size of the neighborhood varies as learning continues, where it starts large and is gradually reduced, making the range of change sharper and sharper.

ordering exists within the output space. The weight vector (reference vector) associated with each output layer neuron is regarded as an exemplar of the kind of input vector to which the neuron will respond. Let the input vector, X, be defined as

$$X = [X_1, X_2, \ldots, X_n]^T$$

and the weight vector, m_i corresponding to output layer neuron i can be written as

$$m_i = [w_{i1}, w_{i2}, \ldots, w_{in}]^T$$

where n is the dimension of the input pattern and $i = 1, 2, \ldots, N$.

Determination of the winning output layer neuron amounts to selecting the output layer neuron whose weight vector m_i best matches the input vector X. Therefore, the input vector is compared with all the m_i to find the smallest Euclidean distance $\|X - m_c\|$ between vectors. If the best match is at the unit with index c, then c is determined by

$$\|X - m_c\| = \min\left\{\|X - m_i\|\right\}$$

or

$$c = \arg\min_i\left\{\|X - m_i\|\right\}$$

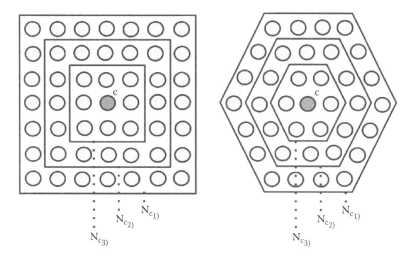

FIGURE 1.22 Rectangular and hexagonal topological neighborhoods N_c of cell c. The radius decreases with time ($t_1 < t_2 < t_3$). The SOFM may be categorized as a nonlinear projection of the probability density function of n-dimensional input data into a one- or two-dimensional lattice of output layer neurons, which comprises the output space such that a meaningful topological ordering exists within the output space. The weight vector (reference vector) associated with each output layer neuron is regarded as an exemplar of the kind of input vector to which the neuron will respond. For the lateral feedback operation, a function is needed that defines the size of the neighborhood surrounding the winning neuron. This function $N_c(t)$ is a function of discrete time (i.e., iteration), as seen in the above figure, where two possible topological neighborhoods are shown. The amount of lateral feedback can be varied over the course of network training. Larger neighborhoods mean more positive feedback, and training takes place at a more global level. It is via a large value of the neighborhood function during the early training of the network that the topological ordering of the network is achieved. Subsequent reduction of the neighborhood then makes the clusters sharper so that the cluster response may be refined.

Thus, X is mapped onto the node c relative to the parameter value m_i (or simply onto m_c).

For the lateral feedback operation, a function is needed that defines the size of the neighborhood surrounding the winning neuron. This function $N_c(t)$ is a function of discrete time (i.e., iteration), as seen in Figure 1.22, where two possible topological neighborhoods are shown. The amount of lateral feedback can be varied over the course of network training. Larger neighborhoods mean more positive feedback, and training takes place at a more global level. It is via a large value of the neighborhood function during the early training of the network that the topological ordering of the network is achieved. Subsequent reduction of the neighborhood then makes the clusters sharper so that the cluster response may be refined. The neighborhood function is utilized to modify the learning process as indicated in Equation 1.5:

$$m_i\left(t+1\right) = \begin{cases} m_i(t) + \alpha(t)[x(t) - m_i(t)] & \forall i \in N_c(t) \\ m_i(t) & \text{otherwise} \end{cases} \tag{1.5}$$

where $\alpha(t)$ is the learning parameter valued at $0 < \alpha(t) < 1$, which decreases with time t.

1.5.2.6 Learning Vector Quantization

Learning vector quantization (Kohonen 1995) is a supervised learning extension of Kohonen network methods (Kohonen 1982). It allows specification of the categories into which inputs will be classified. The designated categories for the training set are known in advance and are part of the training set. The LVQ network architecture has a very similar structure to the SOFM, with an exception that each neuron in the output layer is designated as belonging to one of the several classification categories, as shown in Figure 1.23. In general, several output neurons are assigned to each class. The weight vector (codebook vector) to a given output unit represents an exemplar of the input vectors to which it will most strongly respond. When an input pattern, X, is input to the network, the neuron with the closest (Euclidean distance) codebook vector is declared to be the winner. The training procedure is similar to SOFM, but only the winning neuron is modified in LVQ.

In our own research we have utilized these concepts to produce a two-stage adaptive classification system shown in Figure 1.23, where an SOFM acts as a preprocessor and LVQ as a fine-tuning classifier. Vector quantization (VQ) is concerned with how to divide the input space into disjointed subspaces so that each input vector can be represented by the reproduction vector of the subspace to which it belongs, as shown in Figure 1.24. The collection of possible reproduction vectors is called the codebook of the quantizer. In the combined method, the SOFM algorithm provides an approximate method for computing the codebook vectors in an unsupervised manner, with the approximation being specified by the synaptic weight vectors of the neurons in the feature map. The supervised LVQ algorithm uses class information

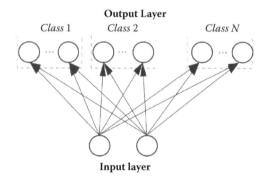

FIGURE 1.23 Learning vector quantizer (LVQ). The figure shows the architecture of the LVQ network. The LVQ network architecture has very similar structure to the SOFM with an exception that each neuron in the output layer is designated as belonging to one of the several classification categories, as shown. In general, several output neurons are assigned to each class. The weight vector to a given output unit represents an exemplar of the input vectors to which it will most strongly respond. When an input pattern is input to the network, the neuron with the closest (Euclidean distance) codebook vector is declared to be the winner. The training procedure is similar to SOFM, but only the winning neuron is modified in LVQ.

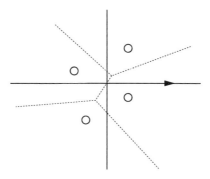

FIGURE 1.24 Adaptive classification system. A simple diagram showing decision boundaries and reproduction vectors for each region is shown. Vector quantization (VQ) is concerned with how to divide the input space into disjointed subspaces so that each input vector can be represented by the reproduction vector of the subspace to which it belongs.

to move the codebook vectors slightly, so as to improve the quality of classifier decision regions.

1.5.2.7   Approaches to Odor Quantification

The quantification of odors or chemicals smelt is a very desirable feature in real life. For example: a type of foodstuff may depend on the concentration of a particular element in a mixture, specific levels of particular chemicals may be identified during environmental monitoring, and the concentration level of a chemical may be critical in the cosmetic and drink industries. The human nose and tongue are employed in many industries as basic tools for determining concentrations of single chemicals or mixtures, where these human senses may vary with the conditions of the human sensory panels at the time of measurement. It is much more difficult to predict concentration levels of single chemicals or mixtures than classification of different chemicals. The properties of single chemicals or mixtures in different concentration levels are similar to each other, and hence the response patterns from the electronic nose system do not show variance as perceived by the human nose. Although each level of a single chemical or mixture can give a unique response pattern, discrimination between these patterns is difficult to achieve.

The neural network approach in pattern recognition enables the automated classification of chemical vapors. However, classification is limited to known classes where a class label is given from the result of recognition. The most widely available multilayer perceptron network, based on back-propagation learning, may be applied to the problems of predicting concentration levels, where the average network output values are given as the result of recognition for each class. However, these networks suffer from local minima problems, and the global optimum is not always guaranteed during the learning process; also, the training process can take a long time. This adds extra burden onto the network architecture due to the complexity of weights adjustment. Radial basis function (RBF) networks have attracted interest due to advantages over multilayer perceptrons, as they are universal approximators but achieve faster learning due to simple architecture, and exhibit none of

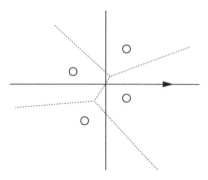

FIGURE 1.25   Radial basis function (RBF) network. The architecture of the RBF network consists of an input layer, a hidden layer, and an output layer. The input vector to the network is passed to the hidden layer nodes via unit connection weights. The hidden layer consists of a set of radial basis functions, the kernel functions. Associated with each hidden layer node is a parameter vector z_j, a center. The hidden layer node calculates the Euclidean distance between the center and the network input vector and then passes the result to a radial basis function. All the radial basis functions in the hidden layer nodes are usually of the same type, and the classifier output is simply a weighted linear summation of the kernel functions.

back-propagation's training pathologies, such as paralysis or local minima problems. A RBF network is a supervised neural network that is often compared with a three-layer feed-forward back-propagation network due to similarity in the network topology; however, its operation is fundamentally different (Figure 1.25). The back-propagation learning algorithm may be viewed as an optimization method known in statistics as stochastic approximation, whereas the RBF network may be viewed as a curve-fitting (approximation) problem in high-dimensional space. The learning of the RBF network is equivalent to finding a surface in multidimensional space that provides a best fit to the training data. In addition, generalization is equivalent to the use of this multidimensional surface to interpolate the test data. RBF networks are universal approximators, i.e., a given network with enough hidden layer neurons that can approximate any continuous function with arbitrary accuracy (Girosi and Poggio 1990; Hartman et al. 1990). However, the main aim of designing an RBF network for practical applications relies on good generalization ability with a minimum number of nodes to reduce unnecessary calculation and processing times.

RBF methods have their origins in techniques that performed exact interpolation of data set points in multidimensional space (Micchelli 1986; Powell 1987). However, those methods were not practical, as the exact interpolation problem requires every input vector to be mapped exactly onto the corresponding target vector. The exact interpolation function is not desired in RBF neural network applications, as it typically forms a highly oscillatory function when there is noise present on the data. The interpolating function that gives the best generalization is one that typically provides much smoother transformation and averages over the noise on the data. In the late 1980s, Broomhead and Lowe (1988) were the first to exploit the use of RBFs in the design of neural networks. Moody and Darken (1989) made a major contribution to the theory, design, and application of the RBF networks. Girosi and Poggio

(1990) emphasized the use of regularization theory applied to the neural network as a method for improved generalization of new data. A network with a finite basis was also developed as a natural approximation of the regularization network. In addition, other extensions, such as moving centers, weighted norm, and networks with different types of basis functions and multiple scales, were also considered.

The modifications introduced for the exact interpolation procedure to obtain the RBF neural network model are as follows: the number of basis functions in the neurons need not be equal to the number of data points; the centers of the basis functions are no longer constrained to be given by the input data vectors, but instead, the determination of suitable centers becomes part of the training process; each basis function is given its own width, whose value is also determined during training; and bias parameters are included in the linear sum that compensate for the difference between the average value of the basis function activations and the corresponding average value of the targets.

The RBF method is one of the possible solutions to real multivariate interpolation problems. A nonlinear function $\Phi(x, z)$, where x is the independent variable and z is the constant parameter, is called a radial basis function when it depends only on the radial distance $r = \|x - z\|$, where z is its "center." For a given m input vectors x_i $\{x_i \in R^n | i = 1, 2, \ldots, m\}$ in n-dimensional space and m real numbers f_i $\{f_i \in R | i = 1, 2, \ldots, m\}$, a function F from R^n to R satisfying the interpolation conditions can be described as $F(x) = f_i$, $i = 1, 2, \ldots, m$. The RBF approach consists of choosing the function F to be an expansion of the form

$$F(x) = \sum_{j=1}^{m} \lambda_j \Phi\left(\|x - z_j\|\right)$$

where the centers of the expansion $z_j = x_j$ must be the known data points, and λ_i $\{\lambda_i \in R | i = 1, 2, \ldots, m\}$ are the corresponding weights.

A variety of approaches for training RBF networks have been developed, where most methods operate in two stages: learning from the centers and widths in the hidden layer and learning from the connection weights of the hidden layer to the output layer (Chen et al. 1992; Freeman and Saad 1995, 1996; Moody and Darken 1989). In these learning algorithms, the network structures are predetermined. Learning algorithms that incorporate structure selection mechanisms have been developed (Poggio and Girosi 1990a, 1990b). Chen et al. (1990, 1991) trained an RBF network using an orthogonal least-square algorithm, which provided a compromise between network performance and network complexity, where the number of hidden layer nodes was automatically determined. A training algorithm proposed by Lee and Rhee (Lee and Kil 1991) was based on a supervised clustering method where the learning started with one hidden layer node with a larger width; additional nodes were created when they were desired, causing changes in the associated widths and locations. The work of Musavi et al. (1992) was also based on a clustering method; however, a larger number of nodes were set at initial learning, which were then merged during processing. The associated widths and locations of the nodes were updated accordingly. Billings and Zheng (1995) proposed genetic algorithms to produce RBF networks that showed an ability to determine appropriate network structures and parameters automatically according to given objective functions. In addition, they claimed that the network had a lower probability of becoming trapped at structural local minima due to the property of the genetic algorithm. It is thought that having an automatic construction of an RBF network during the training process would be useful. However, the efficiency of the RBF algorithm may be lost during practical applications. Sherstinsky and Picard (1996) investigated the efficiency of the orthogonal least-squares (OLS) training for RBF networks and reported that while the OLS method had been believed to find a more efficient selection of the RBF centers than a random-based approach (Chen et al. 1991), such a network did not produce the smallest RBF network for a given approximation accuracy. In practice, the centers are arbitrarily chosen from data points. Apparently such a method cannot guarantee satisfactory performance because it may not satisfy the requirement that centers should suitably sample the input domain. Chen (1995) obtained RBF centers by means of a k-means clustering algorithm, while the network weights were learned using recursive least squares. He also discussed the problem with the conventional k-means clustering algorithm and suggested an enhanced k-means clustering algorithm for the selection of centers.

In our laboratories we have tested many of these algorithms for training and testing RBF networks. If an RBF network is trained with a random value initialization the optimum result is not always guaranteed, so the training may have to be repeated several times before an acceptable recognition level is gained. In addition, the result cannot be reproduced unless the same initial random values are used. Fuzzy c-means algorithms (FCMAs) for locating and initializing centers failed to find optimum centers, especially for higher-dimensional data. The performance of an RBF network can be improved by employing extra centers for each class. The object of the RBF network in practical applications with a minimum number of nodes was to select a center that would well represent the training data set. When the input data have a high dimension, the information becomes more complex, and one center may not be enough to represent the whole class. In the process of assigning multiple centers for a class with the RBF application, optimization with fuzzy c-means algorithms cannot be used as they were designed to choose the optimum cluster center. However, optimization with a learning vector quantization (LVQ) algorithm was shown to be more effective than any of the FCMA techniques. In addition, the iteration time for the LVQ method was several orders of magnitude faster than any of the FCMA methods, especially in the higher-dimensional cases. The combination of the LVQ algorithm and the RBF network was very effective and provided improved results over other methods. The performance of the RBF network depended on the number of centers used for each class. Generally, a better recognition result may be achieved with a higher number of centers, although it also depends on the applications used and input data. However, it is desirable to keep the number of centers as small as possible; the computation required during training is directly related to the complexity of the network. When a large number of the centers are assigned to each class, the

TABLE 1.2

Input Data Preparation for the RBF Network Application for the Quantification of Methanol and Ethanol Mixtures in Fixed Ratios

Class No.	Mixture Ratio Methanol:Ethanol	No. of Patterns Training	Recognition	Database Filename Training	Recognition
1	8:1	11	33	me8et1_t.dba	me8et1.dba
2	4:1	11	33	me4et1_t.dba	me4et1.dba
3	2:1	11	33	me2et1_t.dba	me2et1.dba
4	1:1	11	33	me1et1_t.dba	me1et1.dba
5	1:2	11	33	me1et2_t.dba	me1et2.dba
6	1:4	11	33	me1et4_t.dba	me1et4.dba
7	1:8	11	33	me1et8_t.dba	me1et8.dba

trained network will also have a large number of related parameters. Consequently, a large weight file has to be stored during recognition sessions.

We tested such classifiers with mixtures of two alcohols, methanol and ethanol, with the composition shown in Table 1.2. In experiments with seven methanol and ethanol mixtures, four centers were identified to be the ideal number for each class after considering the computational burden, iteration time, and performance. The RBF application could successfully discriminate between the seven mixtures, and could recognize individual mixtures to the correct quantifying target levels. During experiments, 4 optimized centers were chosen from each training set, which had 11 patterns to train the RBF network. Consequently, the 7 testing data sets, consisting of 33 patterns in each set, were all predicted to the corresponding quantifying targets. The capability of the RBF network was further examined to investigate the response to previously unseen data to the network. Training of the network was carried out with previous data training sets with the same target values as in Table 1.2, but the data set, 1:2 mixture, was omitted from the training sets. Figure 1.26 shows the results, which indicated that—although not perfect—the network was able to interpolate unknown values.

1.6 TOWARD BIOLOGICALLY INSPIRED ARTIFICIAL OLFACTION

All animals continually adapt to changing environments, and plasticity is a characteristic of odor-mediated behavior. This can range from alterations in levels of responsiveness such as sensitization and habituation to more complex forms of associative learning. Computational models of the olfactory system have been developed from a point of view of understanding how the system processes sensory information, and some of these concepts may be applied to artificial olfaction (Davis and Eichenbaum 1991). As described in Section 1.4.2, distributed patterns of activity in response to chemical stimuli are transmitted to the olfactory bulb via olfactory neuron axons that terminate in the glomeruli of its input layer. Electrophysiological

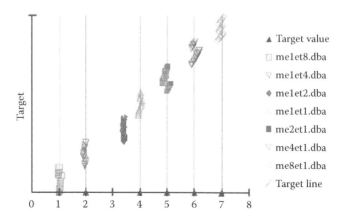

FIGURE 1.26   The figure shows the output display of the RBF application when previously unseen patterns obtained from seven mixtures of analytes were presented. The training data files listed in Table 1.2 were used, with the exception of mixture class 3. One center from each class was used with random initialization. The horizontal axis in the graph indicates the target value or the class index of each mixture associated with the RBF network, and the vertical axis counts the number of patterns. Hence, tested patterns can be interpreted as belonging to the nearest target value or the target line. The recognition output from the RBF network is not the class label but an approximated value. Therefore, the estimated value may be interpreted as the predicted quantification (concentration) value of the chemical being tested. The previously unseen recognition files listed in Table 1.2 were presented—all classes were recognized correctly. The missing exemplar class 3 was also predicted. (Data provided by D.H. Lee, Manchester, UK.)

measurements indicate that the OB filters and transforms these incoming sensory data, so that, for example, signal-to-noise enhancement, normalization, and contrast enhancement operations occur before the processed olfactory information diverges to several different secondary olfactory structures. Many computational models attempt to combine anatomy with function. Signals from ORNs are sent directly to the OB or AL, where they are further processed. OB and AL circuits contain two broad classes of neurons (excitatory projection cells and, for the most part, inhibitory local neurons). Because the mitral and tufted (M/T) cells in mammals have one primary dendrite within one glomerulus or a few glomeruli, and because inhibitory neurons (granule cells) contact nearby M/T cells through their secondary dendrites, this connectivity is often interpreted as underlying a form of lateral inhibition to sharpen M/T cell tuning working in the concept of the Kohonen self-organizing maps described earlier. Schild (1988) elegantly brought together many essential concepts of the olfactory system from a biophysical basis and described a practical model of the stimulus responses of all receptor cells by the use of vector spaces. In this case the morphological convergence pattern between receptor cells and glomeruli is given in the same vector space as the receptor cell activities. It is concluded that sets of mitral cells encoding similar odors work very much in the way of mutually inhibited matched filters. The approach is relatively easy to implement in terms of artificial nose concepts, using the self-organizing map principles described earlier.

Laurent (1999) approached olfactory coding from a systems point of view. It is considered that information is encoded by assemblies of neurons, and that inhibition regulates the global dynamics of these assemblies. Again, while differing in detail, conceptually these ideas converge to similar computational concepts.

Cleland and Linster (2005) discuss in detail many of the challenging computational aspects of the olfactory system and methods that are currently used to model these functions. Without going into detail, several computational models of the OB have suggested that the temporal pattern of spiking among mitral cells may play a role in odor representation. Spiking models have been reviewed and investigated (Brette et al. 2007) and appropriate simulation tools have been selected. Dynamic oscillatory activity patterns are observed in the bulb in response to odor stimulation, and it has been suggested that odor quality may be encoded in terms of dynamic attractors formed in the OB. Models of these phenomena focus on coupled oscillator circuits with feedback occurring between different connections. The dynamics of the olfactory bulb are also tightly coupled to mutual feedback from the higher processing centers in the piriform cortex. Although some models exist that would allow for associative memory to exist in the olfactory bulb so that certain patterns associated with a range of odors become embedded in the circuitry of the olfactory bulb, most models attribute associative memory functions to the piriform cortex. This area of the brain has afferent inputs from the mitral cells of the olfactory bulb, and the neural circuitry is thought to mediate context-dependent learning of odor patterns.

Kauer and White (Kauer and White 1998, 2003; White and Kauer 1999) used a fiber-optic array of sensors and designed a spiking neuron model of the peripheral olfactory system to process signals. In their model, the response of each sensor is converted into a pattern of spikes across a population of ORNs, which then projects to a unique mitral cell. Different odors produce unique spatiotemporal activation patterns across mitral cells, which are then discriminated with a delay line neural network (DLNN). Their model encodes odor quality by the spatial activity across units and odor intensity by the response latency of the units.

Many researchers started out modeling the Hodgkin-Huxley neuron computationally. Such a model is able to exactly reproduce the shape of the action potential of a neuron by taking into account the involved ionic currents, but it is computationally expensive. Several other computationally more efficient models have evolved, and Izhikevich (2003, 2004, 2010) developed a simple model for an artificial neuron that is capable of simulating almost all the functions of a neuron, but is computationally about a hundred times faster than simulation of the Hodgkin-Huxley neuron. It consists of two differential equations with four parameters and accounts for the membrane potential of the neuron, activation of potassium ion currents, and deactivation of sodium ion currents, together with other bias currents. These various models underline the fact that the morphological and biophysical properties of OB neurons and their connections define their computational capabilities.

This concept has been used to implement a cortical-based artificial neural network (CANN) architecture design inspired by the anatomical structure found in the mammalian cortex (Pioggia et al. 2008). A single CANN of 1000 artificial neurons consisting of 200 inhibitory neurons and 800 excitatory neurons was implemented.

This uses a principle of spike timing synchrony that allows a neuron to be activated in correspondence with synchronous input spikes. The researchers used an electronic nose based on an array of conducting polymers together with a composite array-based e-tongue to measure 60 samples of olive oil, correlating these with human panel descriptors. They tested the performance of the cortical-based artificial neural network, a multilayer perceptron, a Kohonen self-organizing map, and a fuzzy Kohonen self-organizing map, and demonstrated that the CANN outperformed the other techniques in terms of its classification and generalization capability.

Raman et al. (2006b) adopted a chemotopic convergence model, where they considered that the signals from many olfactory neurons converge onto a few neighboring glomerular (GL) cells. An ORN is characterized by a vector of log affinities to odorants or molecules in a given chemical problem space. A self-organizing model of convergence based on the Kohenen map described previously was constructed whereby a large number of inputs from ORNS are projected onto a two-dimensional lattice of GL cells. They assume that ORNS converge onto given GL cells on the basis of their selectivity to different molecules. Simulation experiments with this model using a large number of sensory inputs as well as data from a gas sensor array indicate that each analyte evokes a unique glomerular image with higher activity in those GL cells that receive projections from ORNs with high affinity to that analyte. If the analyte concentration increases, additional ORNs with lower affinity are recruited, resulting in an increased activation level and a larger spread of the analyte-specific loci. They found that if the receptive field was too broad, or conversely too narrow, separability between odorant classes deteriorated. In this work, a homogeneous field width was applied, but Alkasab et al. (1999, 2002) indicate that maximum mutual information is achieved with a heterogeneous population of receptors/sensors. Hence, sensors displaying a nonuniform distribution of receptor field widths may produce rather different characteristics when combined in an array.

It is now possible to integrate many olfactory concepts into electronic hardware. For example, a proposed neuron synapse integrated chip (IC) chip set (Horio et al. 2003) makes it possible to construct a scalable and reconfigurable large-scale chaotic neural network with 10,000 neurons and $10,000^2$ synaptic connections. Bioinspired and neuromorphic designs are emerging. Covington et al. (2007) describe an artificial olfactory mucosa. Koikal et al. (2006, 2007) describe an analog very large scale integration (VLSI) implementation of an adaptive neuromorphic olfaction chip. Guerrero-Rivera and Pearce (2007) described attractor-based pattern classification in a field programmable gate array (FPGA) implementation of the olfactory bulb. It is now also possible to make hybrid systems that incorporate olfactory receptor cells, and a bioelectronic nose based on olfactory sensory neural networks in a culture of cells has been described (Liu et al. 2010a, 2010b; Wu et al. 2009). With advances in gene expression Song and coworkers (2009) reported the expression of a human olfactory receptor in *Escherichia coli* bacteria.

This chapter has attempted to paint with a broad brush the breath and marvels of chemosensory systems, and our puny attempts to emulate what nature does so well. By painstakingly breaking down the complex tasks associated with olfaction, we are surely making progress, not just in the artificial olfaction field, but toward understanding brain function and cognition.

ACKNOWLEDGMENT

Results presented on basis neural networks were carried out as part of PhD research by Dr. Dong-Hung Lee, Manchester, UK.

REFERENCES

Abe, H., S. Kanaya, Y. Takahashi, and S. I. Sasaki. 1988. Extended studies of the automated odor-sensing system based on plural semiconductor gas sensors with computerized pattern-recognition techniques. *Analytica Chimica Acta* 215 (1–2): 155–168.

Abe, H., T. Yoshimura, S. Kanaya, Y. Takahashi, Y. Miyashita, and S. I. Sasaki. 1987. Automated odor-sensing system based on plural semiconductor gas sensors and computerized pattern-recognition techniques. *Analytica Chimica Acta* 194: 1–9.

Ache, B. W., and J. M. Young. 2005. Olfaction: Diverse species, conserved principles. *Neuron* 48 (3): 417–430.

Alkasab, T. K., T. C. Bozza, T. A. Cleland, K. M. Dorries, T. C. Pearce, J. White, and J. S. Kauer. 1999. Characterizing complex chemosensors: Information-theoretic analysis of olfactory systems. *Trends in Neurosciences* 22 (3): 102–108.

Alkasab, T. K., J. White, and J. S. Kauer. 2002. A computational system for simulating and analyzing arrays of biological and artificial chemical sensors. *Chemical Senses* 27 (3): 261–275.

Amoore, J. E. 1962a. The stereochemical theory of olfaction. 1. Identification of the seven primary odours. *Proceedings of the Scientific Section, Toilet Goods Association* 37 ((Suppl.)): 1–12.

Amoore, J. E. 1962b. The stereochemical theory of olfaction. 2. Elucidation of the stereochemical properties of the olfactory receptor sites. *Proceedings of the Scientific Section, Toilet Goods Association* 37 ((Suppl.)): 13–23.

Amoore, J. E. 1967. Specific anosmia: A clue to the olfactory code. *Nature* 214: 1095–1098.

Araneda, R. C., A. D. Kini, and S. Firestein. 2000. The molecular receptive range of an odorant receptor. *Nature Neuroscience* 3 (12): 1248–1255.

Arshak, K., E. Moore, G. M. Lyons, J. Harris, and S. Clifford. 2012. A review of gas sensors employed in electronic nose applications. *Sensor Review* 24 (2): 181–198.

Beets, M. J. G. 1978. *Structure-activity relationships in human chemoreception*. London: Applied Science Publishers Ltd.

Benton, R. 2009. Eppendorf winner: Evolution and revolution in odor detection. *Science* 326 (5951): 382–383.

Billings, S. A., and G. L. Zheng. 1995. Radial basis function network configuration using genetic algorithms. *Neural Networks* 8 (6): 877–890.

Boelens, H. 1974. Relationship between the chemical structure of compounds and their olfactive properties. *Cosmetics and Perfumery* 89: 1–7.

Bona, E., R. S. dos Santos Ferreira da Silva, D. Borsato, and D. G. Bassoli. 2012. Self-organizing maps as a chemometric tool for aromatic pattern recognition of soluble coffee. *Acta Scientiarum—Technology* 34 (1): 111–119.

Bott, B., and T. A. Jones. 1986. The use of multisensor systems in monitoring hazardous atmospheres. *Sensors and Actuators* 9 (1): 19–25.

Brette, R., M. Rudolph, T. Carnevale, M. Hines, D. Beeman, J. M. Bower, M. Diesmann, A. Morrison, P. H. Goodman, F. C. Harris, M. Zirpe, T. Natschlaeger, D. Pecevski, B. Ermentrout, M. Djurfeldt, A. Lansner, O. Rochel, T. Vieville, E. Muller, A. P. Davison, S. El Boustani, and A. Destexhe. 2007. Simulation of networks of spiking neurons: A review of tools and strategies. *Journal of Computational Neuroscience* 23 (3): 349–398.

Broomhead, D., and D. Lowe. 1988. Multivariable functional interpolation and adaptive network. *Complex Systems* 2: 321–355.

Buck, L., and R. Axel. 1991. A novel multigene family may encode odorant receptors—A molecular-basis for odor recognition. *Cell* 65 (1): 175–187.

Buck, L. B. 1997a. Information coding in the olfactory system. *Journal of Neurochemistry* 69: S210.

Buck, L. B. 1997b. Molecular mechanisms of odor and pheromone detection in mammals. *Molecular Biology of the Cell* 8: 739.

Buck, T., F. Allen, and M. Dalton. 1965. Detection of chemical species by surface effects on metals and semiconductors. In *Surface effects in detection*, ed. J. I. Bregman and A. Dravnieks, 1–27. Washington, DC: Spartan Books.

Callegari, P., J. Rouault, and P. Laffort. 1997. Olfactory quality: From descriptor profiles to similarities. *Chemical Senses* 22 (1): 1–8.

Chen, S. 1995. Nonlinear time series modelling and prediction using Gaussian RBF networks with enhanced clustering and RLS learning. *Electronics Letters* 31 (2): 117–118.

Chen, S., S. A. Billings, C. F. N. Cowan, and P. M. Grant. 1990. Practical identification of Narmax models using radial basis functions. *International Journal of Control* 52 (6): 1327–1350.

Chen, S., S. A. Billings, and P. M. Grant. 1992. Recursive hybrid algorithm for nonlinear-system identification using radial basis function networks. *International Journal of Control* 55 (5): 1051–1070.

Chen, S., C. F. N. Cowan, and P. M. Grant. 1991. Orthogonal least squares learning algorithm for radial basis function networks. *IEEE Transactions on Neural Networks* 2 (2): 302–309.

Chess, A., L. Buck, M. M. Dowling, R. Axel, and J. Ngai. 1992. Molecular-biology of smell—Expression of the multigene family encoding putative odorant receptors. *Cold Spring Harbor Symposia on Quantitative Biology* 57: 505–516.

Cleland, T. A., and C. Linster. 2005. Computation in the olfactory system. *Chemical Senses* 30 (9): 801–813.

Clyne, P. J., C. G. Warr, M. R. Freeman, D. Lessing, J. H. Kim, and J. R. Carlson. 1999. A novel family of divergent seven-transmembrane proteins: Candidate odorant receptors in *Drosophila*. *Neuron* 22 (2): 327–338.

Covington, J. A., J. W. Gardner, A. Hamilton, T. C. Pearce, and S. L. Tan. 2007. Towards a truly biomimetic olfactory microsystem: An artificial olfactory mucosa. *IET Nanobiotechnology* 1 (2): 15–21.

Crasto, C. J. 2009. Computational biology of olfactory receptors. *Current Bioinformatics* 4 (1): 8–15.

Davis, J. L., and H. Eichenbaum. 1991. *Olfaction: A model system for computational neuroscience*. Cambridge, MA: MIT Press.

Di Natale, C., F. Davide, and A. Damico. 1995a. Pattern-recognition in gas-sensing—Well-stated techniques and advances. *Sensors and Actuators B—Chemical* 23 (2–3): 111–118.

Di Natale, C., F. A. M. Davide, and A. Damico. 1995b. A self-organizing system for pattern-classification—Time-varying statistics and sensor drift effects. *Sensors and Actuators B—Chemical* 27 (1–3): 237–241.

Distante, C., A. Anglani, P. Siciliano, and L. Vasanelli. 2000. mSOM: A SOM based architecture to track dynamic clusters. *Sensors and Microsystems* 270–273.

Distante, C., P. Siciliano, and K. C. Persaud. 2002. Dynamic cluster recognition with multiple self-organising maps. *Pattern Analysis and Applications* 5 (3): 306–315.

Dravnieks, A., and P. J. Trotter. 1965. Polar vapour detection based on thermal modulation of contact potential. *Journal of Scientific Instruments* 42: 624.

Duchamp-Viret, P., A. Duchamp, and M. Vigouroux. 1989. Amplifying role of convergence in olfactory system: A comparative study of receptor cell and second-order neuron sensitivities. *Journal of Neurophysiology* 61 (5): 1085–1094.

Ema, K., M. Yokoyama, T. Nakamoto, and T. Moriizumi. 1989. Odor-sensing system using a quartz-resonator sensor array and neural-network pattern-recognition. *Sensors and Actuators* 18 (3–4): 291–296.

Firestein, S. 2001. How the olfactory system makes sense of scents. *Nature* 413 (6852): 211–218.

Foret, S., and Maleszka, R. 2006. Function and evolution of a gene family encoding odorant binding-like proteins in a social insect, the honey bee (Apis mellifera). *Genome Research* 16 (11): 1404–1413.

Freeman, J. A. S., and D. Saad. 1995. Learning and generalization in radial basis function networks. *Neural Computation* 7 (5): 1000–1020.

Freeman, J. A. S., and D. Saad. 1996. Radial basis function networks: Generalization in over-realizable and unrealizable scenarios. *Neural Networks* 9 (9): 1521–1529.

Gall, M., and R. Müller. 1989. Investigation of gas-mixtures with different MOS gas sensors with regard to pattern-recognition. *Sensors and Actuators* 17 (3–4): 583–586.

Gao, Q., and A. Chess. 1999. Identification of candidate *Drosophila* olfactory receptors from genomic DNA sequence. *Genomics* 60 (1): 31–39.

Garcia, A., A. Aleixandre, J. Gutierrez, and M. C. Horrillo. 2006. Electronic nose for wine discrimination. *Sensors and Actuators B—Chemical* 113 (2): 911–916.

Gardner, J. W. 1991. Detection of vapors and odors from a multisensor array using pattern-recognition. 1. Principal component and cluster-analysis. *Sensors and Actuators B—Chemical* 4 (1–2): 109–115.

Gardner, J. W. 1995. Intelligent gas-sensing using an integrated sensor pair. *Sensors and Actuators B—Chemical* 27 (1–3): 261–266.

Gardner, J. W., E. L. Hines, and H. C. Tang. 1992a. Detection of vapors and odors form a multisensor array using pattern-recognition techniques. 2. Artificial neural networks. *Sensors and Actuators B—Chemical* 9 (1): 9–15.

Gardner, J. W., E. L. Hines, and M. Wilkinson. 1990. Application of artificial neural networks to an electronic olfactory system. *Measurement Science and Technology* 1 (5): 446–451.

Gardner, J. W., T. C. Pearce, S. Friel, P. N. Bartlett, and N. Blair. 1994. A multisensor system for beer flavor monitoring using an array of conducting polymers and predictive classifiers. *Sensors and Actuators B—Chemical* 18 (1–3): 240–243.

Gardner, J. W., H. V. Shurmer, and T. T. Tan. 1992b. Application of an electronic nose to the discrimination of coffees. *Sensors and Actuators B—Chemical* 6 (1–3): 71–75.

Gelis, L., S. Wolf, H. Hatt, E. M. Neuhaus, and K. Gerwert. 2012. Prediction of a ligand-binding niche within a human olfactory receptor by combining site-directed mutagenesis with dynamic homology modeling. *Angewandte Chemie—International Edition* 51 (5): 1274–1278.

GholamHosseini, H., D. Luo, H. Liu, and G. Xu. 2007. Intelligent processing of E-nose information for fish freshness assessment. In *Proceedings of the 2007 International Conference on Intelligent Sensors, Sensor Networks and Information Processing*, 173–177.

Girosi, F., and T. Poggio. 1990. Networks and the best approximation property. *Biological Cybernetics* 63 (3): 169–176.

Guerrero-Rivera, R., and T. C. Pearce. 2007. Attractor-based pattern classification in a spiking FPGA implementation of the olfactory bulb. In *Proceedings of Neural Engineering, 2007. CNE '07. 3rd International IEEE/EMBS Conference on Neural Engineering.* DOI: 10.1109/CNE.2007.369742.

Guo, S., and J. Kim. 2010. Dissecting the molecular mechanism of *Drosophila* odorant receptors through activity modeling and comparative analysis. *Proteins—Structure Function and Bioinformatics* 78 (2): 381–399.

Hansch, C., and T. Fujita. 1964. Rho-sigma-pi analysis. Method for correlation of biological activity + chemical structure. *Journal of the American Chemical Society* 86 (8): 1616.

Hartman, E. J., J. D. Keeler, and J. M. Kowalski. 1990. Layered neural networks with Gaussian hidden units as universal approximations. *Neural Computation* 2 (2): 210–215.

Hekmat-Scafe, D.S., Scafe, C.R., McKinney, A.J., and Tanouye, M.A. 2002. Genome-wide analysis of the odorant-binding protein gene family in Drosophila melanogaster. *Genome Research* 12 (9): 1357–1369.

Hildebrand, J. G. 2001. From molecule to perception: Five hundred million years of olfaction. *Biology International* 41: 41–52.

Hildebrand, J. G., and G. M. Shepherd. 1997. Mechanisms of olfactory discrimination: Converging evidence for common principles across phyla. *Annual Review of Neuroscience* 20: 595–631.

Hines, E. L., and J. W. Gardner. 1994. An artificial neural emulator for an odor sensor array. *Sensors and Actuators B—Chemical* 19 (1–3): 661–664.

Hoare, D. J., J. Humble, D. Jin, N. Gilding, R. Petersen, M. Cobb, and C. McCrohan. 2011. Modeling peripheral olfactory coding in *Drosophila* larvae. *PLoS One* 6 (8).

Horio, Y., K. Aihara, and O. Yamamoto. 2003. Neuron-synapse IC chip-set for large-scale chaotic neural networks. *IEEE Transactions on Neural Networks* 14 (5): 1393–1404.

Imai, T., H. Sakano, and L. B. Vosshall. 2010. Topographic mapping—The olfactory system. *Cold Spring Harbor Perspectives in Biology* 2 (8).

Izhikevich, E. M. 2003. Simple model of spiking neurons. *IEEE Transactions on Neural Networks* 14 (6): 1569–1572.

Izhikevich, E. M. 2004. Which model to use for cortical spiking neurons? *IEEE Transactions on Neural Networks* 15 (5): 1063–1070.

Izhikevich, E. M. 2010. Hybrid spiking models. *Philosophical Transactions of the Royal Society A—Mathematical Physical and Engineering Sciences* 368 (1930): 5061–5070.

Kaissling, K. E. 2009. Olfactory perireceptor and receptor events in moths: A kinetic model revised. *Journal of Comparative Physiology A—Neuroethology Sensory Neural and Behavioral Physiology* 195 (10): 895–922.

Kaissling, K. E., and J. P. Rospars. 2004. Dose-response relationships in an olfactory flux detector model revisited. *Chemical Senses* 29 (6): 529–531.

Kaneyasu, M., A. Ikegami, H. Arima, and S. Iwanaga. 1987. Smell identification using a thick-film hybrid gas sensor. *IEEE Transactions on Components Hybrids and Manufacturing Technology* 10 (2): 267–273.

Kangas, J. A., T. K. Kohonen, and J. T. Laaksonen. 1990. Variants of self-organizing maps. *IEEE Transactions on Neural Networks* 1 (1): 93–99.

Karhunen, J., and J. Joutsensalo. 1995. Generalizations of principal component analysis, optimization problems, and neural networks. *Neural Networks* 8 (4): 549–562.

Kauer, J., and J. White. 1998. A portable artificial nose based on olfactory principles. *Society for Neuroscience Abstracts* 24 (1–2): 652.

Kauer, J. S., and J. White. 2001. Imaging and coding in the olfactory system. *Annual Review of Neuroscience* 24: 963–979.

Kauer, J. S., and J. White. 2003. Representation of odor information in the olfactory system: From biology to an artificial nose. *Sensors and Sensing in Biology and Engineering* 305–322.

Kay, L. M., and M. Stopfer. 2006. Information processing in the olfactory systems of insects and vertebrates. *Seminars in Cell and Developmental Biology* 17: 433–442.

Kim, M.S., Repp, A., and Smith, D.P. 1998. LUSH odorant-binding protein mediates chemosensory responses to alcohols in Drosophila melanogaster. *Genetics* 150 (2): 711–721.

Kodogiannis, V. S., J. N. Lygouras, A. Tarczynski, and H. S. Chowdrey. 2008. Artificial odor discrimination system using electronic nose and neural networks for the identification of urinary tract infection. *IEEE Transactions on Information Technology in Biomedicine* 12 (6): 707–713.

Kohonen, T. 1982. Self-organized formation of topologically correct feature maps. *Biological Cybernetics* 43 (1): 59–69.

Kohonen, T. 1995. Learning vector quantization. In *The handbook of brain theory and neural networks*, ed. M. A. Arbib, 537–540. Cambridge, MA: MIT Press.

Koickal, T. J., A. Hamilton, T. C. Pearce, S. L. Tan, J. A. Covington, and J. W. Gardner. 2006. *Analog VLSI design of an adaptive neuromorphic chip for olfactory systems.*

Koickal, T. J., A. Hamilton, S. L. Tan, J. A. Covington, J. W. Gardner, and T. C. Pearce. 2007. Analog VLSI circuit implementation. Of an adaptive neuromorphic olfaction chip. *IEEE Transactions on Circuits and Systems I—Regular Papers* 54 (1): 60–73.

Korsching, S. I. 2001. Odor maps in the brain: Spatial aspects of odor representation in sensory surface and olfactory bulb. *Cellular and Molecular Life Sciences* 58 (4): 520–530.

Kowalski, B. R., and C. F. Bender. 1972. Pattern-recognition—Powerful approach to interpreting chemical data. *Journal of the American Chemical Society* 94 (16): 5632.

Kowalski, B. R., and C. F. Bender. 1973. Pattern-recognition. 2. Linear and nonlinear methods for displaying chemical data. *Journal of the American Chemical Society* 95 (3): 686–693.

Laurent, G. 1999. A systems perspective on early olfactory coding. *Science* 286 (5440): 723–728.

Lee, S., and R. M. Kil. 1991. A Gaussian potential function network with hierarchically self-organizing learning. *Neural Networks* 4 (2): 207–224.

Leinwand, S. G., and S. H. Chalasani. 2011. Olfactory networks: From sensation to perception. *Current Opinion in Genetics and Development* 21 (6): 806–811.

Liu, Q., W. Ye, L. Xiao, L. Du, N. Hu, and P. Wang. 2010b. Extracellular potentials recording in intact olfactory epithelium by microelectrode array for a bioelectronic nose. *Biosensors and Bioelectronics* 25 (10): 2212–2217.

Liu, Q. J., W. W. Ye, N. Hu, H. Cai, H. Yu, and P. Wang. 2010a. Olfactory receptor cells respond to odors in a tissue and semiconductor hybrid neuron chip. *Biosensors and Bioelectronics* 26 (4): 1672–1678.

Lozano, J., T. Arroyo, J. Santos, J. Cabellos, and M. Horrillo. 2008. Electronic nose for wine ageing detection. *Sensors and Actuators B—Chemical* 133 (1): 180–186.

Malnic, B., J. Hirono, T. Sato, and L. B. Buck. 1999. Combinatorial receptor codes for odors. *Cell* 96 (5): 713–723.

Martin, J. P., A. Beyerlein, A. M. Dacks, C. E. Reisenman, J. A. Riffell, H. Lei, and J. G. Hildebrand. 2011. The neurobiology of insect olfaction: Sensory processing in a comparative context. *Progress in Neurobiology* 95 (3): 427–447.

Micchelli, C. A. 1986. Interpolation of scattered data—Distance matrices and conditionally positive definite functions. *Constructive Approximation* 2 (1): 11–22.

Minor, A.V., and Kaissling, K.E. 2003. Cell responses to single pheromone molecules may reflect the activation kinetics of olfactory receptor molecules. *Journal of Comparative Physiology A: Neuroethology Sensory Neural and Behavioral Physiology* 189 (3): 221–230.

Moens, M., A. Smet, B. Naudts, J. Verhoeven, M. Ieven, P. Jorens, H. J. Geise, and F. Blockhuys. 2006. Fast identification of ten clinically important micro-organisms using an electronic nose. *Letters in Applied Microbiology* 42 (2): 121–126.

Mombaerts, P., F. Wang, C. Dulac, S. K. Chao, A. Nemes, M. Mendelsohn, J. Edmondson, and R. Axel. 1996b. Visualizing an olfactory sensory map. *Cell* 87 (4): 675–686.

Mombaerts, P., F. Wang, C. Dulac, R. Vassar, S. K. Chao, A. Nemes, M. Mendelsohn, J. Edmondson, and R. Axel. 1996a. The molecular biology of olfactory perception. *Cold Spring Harbor Symposia on Quantitative Biology* 61: 135–145.

Moncrieff, R. 1961. Instrument for measuring and classifying odors. *Journal of Applied Physiology* 16 (4): 742.

Moody, J., and C. J. Darken. 1989. Fast learning in networks of locally-tuned processing units. *Neural Computation* 1 (2): 281–294.

Mori, K., H. Nagao, and Y. Yoshihara. 1999. The olfactory bulb: Coding and processing of odor molecule information. *Science* 286 (5440): 711–715.

Moriizumi, T., T. Nakamoto, and Y. Sakuraba. 1991. Pattern recognition in electronic noses by artificial neural network methods. In *Sensors and sensory systems for an electronic nose*, ed. J. W. Gardner and P. N. Bartlett, 217–236. Nat Science Series E: Volume 212. Dordrecht, The Netherlands: Kluwer Academic Publishers.

Müller, R., and E. Lange. 1986. Multidimensional sensor for gas-analysis. *Sensors and Actuators* 9 (1): 39–48.

Murlis, J., J. S. Elkinton, and R. T. Carde. 1992. Odor plumes and how insects use them. *Annual Review of Entomology* 37: 505–532.

Murlis, J., and C. D. Jones. 1981. Fine-scale structure of odor plumes in relation to insect orientation to distant pheromone and other attractant sources. *Physiological Entomology* 6 (1): 71–86.

Musavi, M. T., W. Ahmed, K. H. Chan, K. B. Faris, and D. M. Hummels. 1992. On the training of radial basis function classifiers. *Neural Networks* 5 (4): 595–603.

Nakamoto, T., A. Fukuda, T. Moriizumi, and Y. Asakura. 1991. Improvement of identification capability in an odor-sensing system. *Sensors and Actuators B—Chemical* 3 (3): 221–226.

Nakamoto, T., K. Fukunishi, and T. Moriizumi. 1990. Identification capability of odor sensor using quartz-resonator array and neural-network pattern-recognition. *Sensors and Actuators B—Chemical* 1 (1–6): 473–476.

Pelosi, P. 2001. The role of perireceptor events in vertebrate olfaction. *Cellular and Molecular Life Sciences* 58 (4): 503–509.

Pelosi, P., and Maida, R. 1990. Odorant-binding proteins in vertebrates and insects: Similarities and possible common function. *Chemical Senses* 15 (2): 205–215.

Pelosi, P., and Maida, R. 1995. The physiological functions of odorant-binding proteins. *Biophysics* (English translation of *Biofizika*) 40 (1): 143–151.

Penza, M., and G. Cassano. 2003. Application of principal component analysis and artificial neural networks to recognize the individual VOCs of methanol/2-propanol in a binary mixture by SAW multi-sensor array. *Sensors and Actuators B—Chemical* 89 (3): 269–284.

Penza, M., G. Cassano, and F. Tortorella. 2002. Identification and quantification of individual volatile organic compounds in a binary mixture by SAW multisensor array and pattern recognition analysis. *Measurement Science and Technology* 13 (6): 846–858.

Perez-Orive, J., O. Mazor, G. C. Turner, S. Cassenaer, R. I. Wilson, and G. Laurent. 2002. Oscillations and sparsening of odors representations in the mushroom body. *Science* 297: 359–365.

Persaud, K. C. 2005. Polymers for chemical sensing. *Materials Today* 8 (4): 38–44.

Persaud, K., and G. Dodd. 1982. Analysis of discrimination mechanisms in the mammalian olfactory system using a model nose. *Nature* 299 (5881): 352–355.

Persaud, K. C., and P. Pelosi. 1985. An approach to an artificial nose. *Transactions of the American Society for Artificial Internal Organs* 31: 297–300.

Persaud, K. C., and P. Pelosi. 1992. Sensor arrays using conducting polymers for an artificial nose. In *Sensors and sensory systems for an electronic nose*, ed. J. W. Gardner and P. N. Bartlett, 237–256. Berlin: Springer-Verlag.

Pilpel, Y., and D. Lancet. 1999. The variable and conserved interfaces of modeled olfactory receptor proteins. *Protein Science* 8 (5): 969–977.

Pioggia, G., M. Ferro, F. Di Francesco, A. Ahluwalia, and D. De Rossi. 2008. Assessment of bioinspired models for pattern recognition in biomimetic systems. *Bioinspiration and Biomimetics* 3 (1).

Poggio, T., and F. Girosi. 1990a. Networks for approximation and learning. *Proceedings of the IEEE* 78 (9): 1481–1497.

Poggio, T., and F. Girosi. 1990b. Regularization algorithms for learning that are equivalent to multilayer networks. *Science* 247 (4945): 978–982.

Polak, E. H. 1973. Multiple profile-multiple receptor site model for vertebrate olfaction. *Journal of Theoretical Biology* 40 (3): 469–484.

Poo, C., and J. S. Isaacson. 2009. Odor representations in olfactory cortex: "Sparse" coding, global inhibition, and oscillations. *Neuron* 62 (6): 850–861.

Powell, M. J. D. 1987. Radial basis functions for multivariable interpolation: A review. In *Algorithms for approximation*, ed. J. C. Mason and M. G. Cox, 143–167. New York: Clarendon Press.

Raman, B., P. A. Sun, A. Gutiérrez-Gálvez, and R. Gutiérrez-Osuna. 2006a. Processing of chemical sensor arrays with a biologically inspired model of olfactory coding. *IEEE Transactions on Neural Networks* 17 (4): 1015–1024.

Raman, B., P. A. Sun, A. Gutiérrez-Gálvez, and R. Gutierrez-Osuna. 2006b. Processing of chemical sensor arrays with a biologically inspired model of olfactory coding. *IEEE Transactions on Neural Networks* 17 (4): 1015–1024.

Reinhard, J., and M. V. Srinivasanand. 2009. The role of scents in honey bee foraging and recruitment. In *Food exploitation by social insects: Ecological, behavioral, and theoretical approaches*, ed. S. Jarau and M. Hrncir, 165–182. Boca Raton, FL: CRC Press.

Ressler, K. J. 1994. Information coding in the olfactory system—Evidence for a stereotyped and highly organized epitope map in the olfactory-bulb. *Cell* 79 (7): 1245–1255.

Ressler, K. J., S. L. Sullivan, and L. B. Buck. 1993. A zonal organization of odorant receptor gene-expression in the olfactory epithelium. *Cell* 73 (3): 597–609.

Ressler, K. J., S. L. Sullivan, and L. B. Buck. 1994. Information coding in the olfactory system: Evidence for a stereotyped and highly organized epitope map in the olfactory bulb. *Cell* 79 (7): 1245–1255.

Rospars, J. P., P. Lucas, and M. Coppey. 2007. Modelling the early steps of transduction in insect olfactory receptor neurons. *Biosystems* 89 (1–3): 101–109.

Rumelhart, D. E., G. E. Hinton, and R. J. Williams. 1986. Learning representations by back-propagating errors. *Nature* 323 (6088): 533–536.

Rupe, H., and K. von Majewski. 1900. Notizen. *Berichte der deutschen chemischen Gesellschaft* 33: 3401–3410.

Sammon, J. W. 1969. A nonlinear mapping for data structure analysis. *IEEE Transactions on Computers* C 18 (5): 401.

Sammon, J. W. 1970. Interactive pattern analysis and classification. *IEEE Transactions on Computers* C 19 (7): 594.

Sanz, G., C. Schlegel, J. C. Pernollet, and L. Briand. 2005. Comparison of odorant specificity of two human olfactory receptors from different phylogenetic classes and evidence for antagonism. *Chemical Senses* 30 (1): 69–80.

Schild, D. 1988. Principles of odor coding and a neural network for odor discrimination. *Biophysical Journal* 54 (6): 1001–1011.

Schoenfeld, T. A., and T. A. Cleland. 2006. Anatomical contributions to odorant sampling and representation in rodents: Zoning in on sniffing behavior. *Chemical Senses* 31 (2): 131–144.

Sell, C. 2006. On the unpredictability of odor. *Angewandte Chemie—International Edition* 45 (38): 6254–6261.

Sengupta, P., J. H. Chou, and C. I. Bargmann. 1996. odr-10 encodes a seven transmembrane domain olfactory receptor required for responses to the odorant diacetyl. *Cell* 84 (6): 899–909.

Shepherd, G. M. 1990. In *Frank Allison Linville's R. H. Wright lectures on olfactory research*, ed. K. Colbow, 61–109. Burnaby, British Columbia, Canada: Simon Fraser University.

Shepherd, G. M., W. R. Chen, and C. A. Greer. 2004. Olfactory bulb. In *Synaptic organization of the brain*, ed. G. M. Shepherd, 165–216. New York: Oxford University Press.

Sherstinsky, A., and R. W. Picard. 1996. On the efficiency of the orthogonal least squares training method for radial basis function networks. *IEEE Transactions on Neural Networks*
7 (1): 195–200.

Shurmer, H. V., and J. W. Gardner. 1992. Odor discrimination with an electronic nose. *Sensors
and Actuators B—Chemical* 8 (1): 1–11.

Shurmer, H. V., J. W. Gardner, and H. T. Chan. 1989. The application of discrimination techniques to alcohols and tobaccos using tin-oxide sensors. *Sensors and Actuators* 18
(3–4): 361–371.

Shurmer, H. V., J. W. Gardner, and P. Corcoran. 1990. Intelligent vapor discrimination using
a composite 12-element sensor array. *Sensors and Actuators B—Chemical* 1 (1–6):
256–260.

Singh, S., E. L. Hines, and J. W. Gardner. 1996. Fuzzy neural computing of coffee and tainted-
water data from an electronic nose. *Sensors and Actuators B—Chemical* 30 (3): 185–190.

Sleight, R. 1990. Evolutionary strategies and learning for neural networks. MSc, UMIST.

Song, H. S., S. H. Lee, E. H. Oh, and T. H. Park. 2009. Expression, solubilization and purifica-
tion of a human olfactory receptor from *Escherichia coli*. *Current Microbiology* 59 (3):
309–314.

Spehr, M., and T. Leinders-Zufall. 2005. One neuron—multiple receptors: Increased complexity in olfactory coding? *Science's STKE: Signal Transduction Knowledge Environment*
2005 (285): e25.

Steinbrecht, R.A. 1998. *Odorant-binding proteins: Expression and function*. Annals of the
New York Academy of Sciences, Olfaction, and Taste XII, pp. 323–332.

Stetter, J. R., M. W. Findlay, G. J. Maclay, J. Zhang, S. Vaihinger, and W. Gopel. 1990. Sensor
array and catalytic filament for chemical-analysis of vapors and mixtures. *Sensors and
Actuators B—Chemical* 1 (1–6): 43–47.

Stetter, J. R., P. C. Jurs, and S. L. Rose. 1986. Detection of hazardous gases and vapors—
Pattern-recognition analysis of data from an electrochemical sensor array. *Analytical
Chemistry* 58 (4): 860–866.

Su, C. Y., K. Menuz, and J. R. Carlson. 2009. Olfactory perception: Receptors, cells, and circuits. *Cell* 139 (1): 45–59.

Sundgren, H., I. Lundstrom, F. Winquist, I. Lukkari, R. Carlsson, and S. Wold. 1990.
Evaluation of a multiple gas-mixture with a simple MOSFET gas sensor array and
pattern-recognition. *Sensors and Actuators B—Chemical* 2 (2): 115–123.

Szyszka, P., M. Ditzen, A. Galkin, C. G. Galizia, and R. Menzel. 2005. Sparsening and temporal sharpening of olfactory representations in the honeybee mushroom bodies. *Journal
of Neurophysiology* 94 (5): 3303–3313.

Tegoni, M., Pelosi, P., Vincent, F., Spinelli, S., Campanacci, V., Grolli, S., Ramoni, R., and
Cambillau, C. 2000. Mammalian odorant binding proteins. *Biochimica et Biophysica
Acta-Protein Structure and Molecular Enzymology* 1482 (1–2): 229–240.

Triller, A., E. A. Boulden, A. Churchill, H. Hatt, J. Englund, M. Spehr, and C. S. Sell. 2008.
Odorant-receptor interactions and odor percept: A chemical perspective. *Chemistry and
Biodiversity* 5 (6): 862–886.

Turin, L., and F. Yoshii. 2003. Structure-odor relations: A modern perspective. In *Handbook of
olfaction and gustation*, ed. R. L. Doty, 457–492. New York: Marcel Dekker.

Vassar, R., S. K. Chao, R. Sitcheran, J. M. Nuñez, L. B. Vosshall, and R. Axel. 1994. Topographic
organization of sensory projections to the olfactory bulb. *Cell* 79 (6): 981–991.

Vassar, R., J. Ngai, and R. Axel. 1993. Spatial segregation of odorant receptor expression in
the mammalian olfactory epithelium. *Cell* 74 (2): 309–318.

Vincent, F., R. Ramoni, S. Spinelli, S. Grolli, M. Tegoni, and C. Cambillau. 2004. Crystal
structures of bovine odorant-binding protein in complex with odorant molecules.
European Journal of Biochemistry 271 (19): 3832–3842.

Vincent, F., S. Spinelli, R. Ramoni, S. Grolli, P. Pelosi, C. Cambillau, and M. Tegoni. 2000. Complexes of porcine odorant binding protein with odorant molecules belonging to different chemical classes. *Journal of Molecular Biology* 300 (1): 127–139.

Vogt, R. 2006. Odorant/pheromone metabolism in insects. *Chemical Senses* 31 (5): A7–A8.

Vogt, R. G., and L. M. Riddiford. 1981. Pheromone binding and inactivation by moth antennae. *Nature* 293 (5828): 161–163.

Vogt, R. G., and L. M. Riddiford. 1984. The biochemical design of pheromone reception— Transport and inactivation. *Chemical Senses* 8 (3): 268.

Wang, F., A. Nemes, M. Mendelsohn, and R. Axel. 1998. Odorant receptors govern the formation of a precise topographic map. *Cell* 93 (1): 47–60.

Weimar, U., K. D. Schierbaum, W. Gopel, and R. Kowalkowski. 1990. Pattern-recognition methods for gas-mixture analysis—Application to sensor arrays based upon SnO2. *Sensors and Actuators B—Chemical* 1 (1–6): 93–96.

Weyerstahl, P. 1994. Odor and structure. *Journal fur Praktische Chemie-Chemiker-Zeitung* 336 (2): 95–109.

White, J., and J. S. Kauer. 1999. Odor recognition in an artificial nose by spatio-temporal processing using an olfactory neuronal network. *Neurocomputing* 26–27: 919–924.

Wilkens, W. F., and J. D. Hartman. 1964. Electronic analog for olfactory processes. *Annals of the New York Academy of Sciences* 116 (A2): 608.

Wilson, R. I., and Z. F. Mainen. 2006. Early events in olfactory processing. *Annual Review of Neuroscience* 29 (1): 163–201.

Wold, S., K. Esbensen, and P. Geladi. 1987. Principal component analysis. *Chemometrics and Intelligent Laboratory Systems* 2 (1–3): 37–52.

Wu, C. S., P. H. Chen, Q. Yuan, and P. Wang. 2009. Response enhancement of olfactory sensory neurons-based biosensors for odorant detection. *Journal of Zhejiang University— Science B* 10 (4): 285–290.

Xu, P. X. 2005. Eppendorf 2005 Winner—A *Drosophila* OBP required for pheromone signaling. *Science* 310 (5749): 798–799.

Xu, Y. L., P. He, L. Zhang, S. Q. Fang, S. L. Dong, Y. J. Zhang, and F. Li. 2009. Large-scale identification of odorant-binding proteins and chemosensory proteins from expressed sequence tags in insects. *BMC Genomics* 10.

Yoshii, F., S. Hirono, Q. Liu, and I. Moriguchi. 1992. 3-Dimensional structure model for benzenoid musks expressed by computer-graphics. *Chemical Senses* 17 (5): 573–582.

Yoshii, F., Q. Liu, S. Hirono, and I. Moriguchi. 1991. Quantitative structure-activity-relationships of structurally similar odorless and odoriferous benzenoid musks. *Chemical Senses* 16 (4): 319–328.

Zaromb, S., and J. R. Stetter. 1984. Theoretical basis for identification and measurement of air contaminants using an array of sensors having partly overlapping selectivities. *Sensors and Actuators* 6 (4): 225–243.

Zuppa, M., C. Distante, P. Siciliano, and K. C. Persaud. 2004. Drift counteraction with multiple self-organising maps for an electronic nose. *Sensors and Actuators B—Chemical* 98 (2–3): 305–317.

2 Study of the Coding Efficiency of Populations of Olfactory Receptor Neurons and Olfactory Glomeruli

Agustín Gutiérrez-Gálvez and Santiago Marco

CONTENTS

2.1 INTRODUCTION

The olfactory system has been optimized over evolutionary time to perform an exquisite function: analyze odorant molecules by their molecular features, and synthesize holistic representations of them when presented in complex mixtures. It has been estimated that the olfactory system is able to detect approximately 10,000 odors (Axel 1995) over a large range of concentrations. However, unlike the sense of hearing or vision, this modality has been elusive to psychophysical analysis because no simple set of physical properties, such as light wavelengths for sight or sound frequency for hearing, has been found. Rather, olfaction appears to be intrinsically multidimensional. Along with the multidimensional nature of olfaction, the striking similarity of different olfactory systems across phyla (Hildebrand and Shepherd 1997) suggests that its architecture has been optimized to reflect basic properties of olfactory stimuli.

The objective of this study is to analyze how odor intensity and odor quality information is encoded on the first stages of the olfactory pathway: the olfactory epithelium and the olfactory glomerular layer. To study the olfactory epithelium, we built computational models of olfactory receptor neuron (ORN) populations based on their experimental statistical distributions. These models are based on the detailed characterization of the odor concentration response of ORN populations reported by Rospars et al. (2003). To study the glomerular layer, we modeled the ORN axon projections and lateral inhibitory interactions occurring at the olfactory glomeruli. The odor intensity and odor quality information conveyed by these two stages of the olfactory system is evaluated using the information theory. We consider the amount of information transmitted as a measure of the efficiency of the coding strategy followed at each stage.

In this chapter we first present an introduction that contains background regarding the early stages of the olfactory system, the ORN models used in this work, and a description of the information theoretic measure used: the mutual information. Then, we present the three studies performed in this work. In Section 2.2, we study the coding of odor intensity at the olfactory epithelium. In the next section (2.3), we perform the odor intensity study at the following anatomical stage: the glomerular layer. Finally, in Section 2.4 the study of odor intensity is extended to odor quality as well and applied at the olfactory epithelium.

2.1.1 The Early Olfactory Pathway

We focus our study on the first two stages of the olfactory pathway: the olfactory epithelium and the glomerular layer within the olfactory bulb. In the olfactory epithelium, the molecular properties of the odorants are transduced into electrical signals through a collection of olfactory receptor neurons (ORNs). Mammals have tens of millions of ORNs (Hildebrand and Shepherd 1997; Doty 1991), which belong to as many as 1000 different types of receptors (Ma and Shepherd 2000). The prevailing hypothesis about olfactory primary reception is that ORNs do not respond to specific molecules, but rather to specific molecular features of an odorant molecule, commonly referred to as odotopes (Shepherd 1987, 1994), such as carbon-chain length, the presence of benzene rings, or different functional groups (e.g., ester, aldehydes).

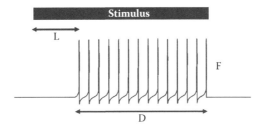

FIGURE 2.1 Three parameters of the ORN output action potentials characterized by Rospars et al.'s (2003) model: latency (L), duration (D), and frequency (F).

Considering that most odorants in the environment consist of mixtures of volatile molecules (e.g., roasted coffee has been estimated to contain the order of 600 volatile components), and that each molecule can contain several odotopes, an odorant is then detected as a large combination of specific odotopes.

ORNs project in a very orderly fashion into spherical regions of neuropil known as glomeruli. Each glomerulus receives axons from one type of ORN, and each ORN type projects into one or a few glomeruli (Vassar 1994; Ressler et al. 1994). Therefore, at the glomerular level, olfactory information can be thought of as being represented by an image of the molecular features of the stimulus. Two types of neurons can be found within the glomerular layer: projection neurons (mitral and tufted cells) and local interneurons (periglomerular and short axon cells) (Shepherd and Greer 2004; Aungst et al. 2003). Projection neurons receive inputs from their main dendrite, which is located at the glomeruli, and transmit this activation to higher areas of the olfactory system. The second type of cells, local interneurons, provides lateral connectivity between projection neurons at two different ranges of the glomerular layer. First, periglomerular cells provide inhibitory interconnection between projection neuron dendrites within glomeruli. Second, short axon cells interconnect tufted cells and periglomerular cells between glomeruli. Even though short axon cells are excitatory, they synapse to periglomerular cells, resulting in an inhibitory effect.

2.1.2 OLFACTORY RECEPTOR NEURON MODEL

The olfactory receptor neuron response models used in this work were proposed by Rospars et al. (2003). These models capture the relationship between ORN output action potentials and odor concentration using three parameters: frequency, latency, and duration of action potentials (Figure 2.1). For each one of these parameters, they find a mathematical expression that best fits the variation of the parameter with odor concentration. This fitting is performed with the extracellular recordings of the spiking activity of receptor neurons in the frog olfactory epithelium when four odorants were presented at precisely controlled concentrations.

2.1.2.1 Response Model for Different Odorant Concentrations

The expressions proposed by Rospars et al. (2003) for the frequency, latency, and duration of the output action potentials are the following:

Frequency:
$$F(C) = F_M \left(\frac{2}{1 + \exp(-(\log_2 10) \cdot n \cdot (C - C_T))} - 1 \right) \tag{2.1}$$

$$F(C < C_T) = 0$$

Latency:
$$L(C) = L_a \exp\left(-\lambda(C - C_T)\right) + L_m \tag{2.2}$$

Duration:
$$D(C) = D_M \frac{C - C_T}{C_M - C_T} \cdot \exp\left(1 - \frac{C - C_T}{C_M - C_T}\right) \tag{2.3}$$

where F_M is the maximum firing frequency, C_T is the threshold concentration, and n controls the slope of the dose-response frequency curve. In Equation 2.2, L_m is the minimum latency, $L_M = L_a + L_m$ is the maximum frequency, and λ is the slope. In Equation 2.3 the maximum duration at the dose C_M is D_M. The concentration is defined as $C = \log_{10} M$, and M is the molarity of the saturated vapor. Figure 2.2 shows the values of the spiking frequency (a), response latency (b), and response duration (c) extracted from Equations 2.1 to 2.3, respectively, as the odor concentration is increased from –7 to –3.

2.1.2.2 Response Model for Different Odorants and Odorant Concentrations

Rospars et al. (2003) also presented an anatomic electrical model that captured the frequency rate of the ORNs in terms of the odorant concentration and number of other structural parameters of the neuron. One of these parameters is the affinity of the ORN to specific ligands that characterizes its response to different odorants. The model accounts for the three main steps of chemotransduction: the change of conductance due to opening of ion channels, the potential of the axon at its initial segment, and the firing frequency. The following three equations correspond to each one of these steps:

Membrane conductance:
$$g(M) = \frac{g_M}{1 + \left(\dfrac{M_{g/2}}{M}\right)^{n_g}} \tag{2.4}$$

Receptor potential:
$$V(g) = E \frac{N_{cil} r_{in} g}{N_{cil} r_{in}(1 + g) + \sqrt{1 + g} \coth\left(\sqrt{1 + g l_{cil}}\right)} \exp\left(-l_{den}\right) \tag{2.5}$$

Firing frequency:
$$F(V) = \begin{cases} 0 & \text{for } V < V_t \\ F_{max} V / E' & \text{for } V_t < V < E' \\ F_{max} & \text{for } V < E' \end{cases} \tag{2.6}$$

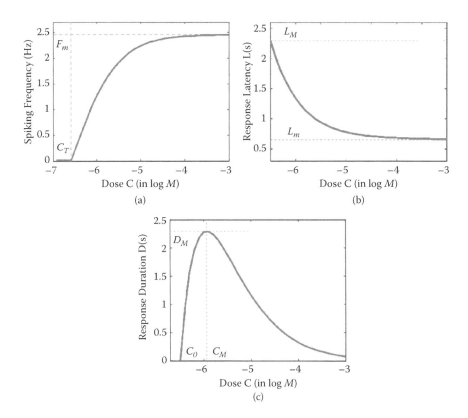

FIGURE 2.2 ORN model response in terms of concentration: (a) spiking frequency, (b) response latency, and (c) response duration.

2.1.3 Mutual Information

Information theory (Shannon and Weaver 1949; Cover and Thomas 2006) provides a theoretical framework to quantifying the ability of a communication channel or coding scheme to convey information. This theory is suitable to evaluate the coding efficiency of a population of ORNs (Rieke et al. 1997). In this study, we are interested in measuring to what extent the amount of information at the input of the ORNs is transmitted to their output. To measure this quantity, information theory provides a magnitude called mutual information (MI), which is defined as

$$MI = \sum_{i=1}^{N_x} \sum_{j=1}^{N_y} p(y_j) p\left(x_i \mid y_j\right) \log_2 \frac{p\left(x_i \mid y_j\right)}{p(x_i)} \tag{2.7}$$

where y_j and x_i are the input and output states, $p(y_i)$ and $p(x_j)$ are their probability, and N_y and N_x are the number of possible input states, respectively.

2.2   STUDY OF ODOR INTENSITY CODING IN AN ORN POPULATION

In this section we estimate the information capacity of a synthetically generated population of olfactory receptor neurons (ORN) expressing the same olfactory receptor protein (ORP). This study is aimed at improving our understanding of odor intensity encoding at the olfactory epithelium. The ORNs of the olfactory epithelium belong to different types depending on the ORP they express (Buck and Axel 1991). Different ORPs allow binding to different odorant molecules or molecular features. Consequently, the quality of the odorants is captured at the olfactory epithelium as a population code across ORNs of different types. At the same time, the intensity of the odorants is thought to be mainly encoded by the firing frequency of the ORNs (Hildebrand and Shepherd 1997).

However, the apparent simplicity of this intensity coding scheme has been challenged by recent experimental results. Grosmaitre et al. (2006) have reported that ORNs expressing the same odorant receptor (MOR23) do not respond in the same way to the same amount of odorant. Even though all ORNs expressing MOR23 respond to the same odorant, each one of them has a different dose-response curve. The dose-response curve reflects the activity of each ORN in terms of the concentration of odorant used to elicit that activity. This cellular heterogeneity suggests that odor intensity is encoded as a population code across ORNs of the same type.

In this study we propose a computational approach to study the advantages of heterogeneous ORN populations, as opposed to homogeneous ones, to encode for odor intensity. We have generated a population of ORN mimicking the existent diversity on a population of ORN of the same type. The population has been generated following the statistical distributions experimentally found by Rospars et al. (2003) for the different parameters of the ORN: firing frequency, firing threshold, maximum frequency, response duration, and latency (Equations 2.1 to 2.3). To understand the role that each one of these parameters play on the odor intensity coding process, we analyze the information conveyed by the population as the statistical distribution of these parameters is varied.

We have performed this study for two types of ORN models: static and dynamic. Since the firing frequency is supposed to capture most of the odorant information, we built first a static model that considered only the firing frequency of the ORN and not time-dependent parameters. Second, we extended it by building a dynamic model of individual ORNs, which considers not only the firing frequency but also the latency, and the duration of the spike train. This will allow our study to evaluate the role of the time on odor intensity encoding.

2.2.1   STATIC ORN POPULATIONS

An interesting question that arises about the encoding ability of any neuron population is the following: What is the coding advantage of using a population with different neuron types (inhomogeneous population), as opposed to encoding using one with the same neuron type (homogeneous population)? With the aim of addressing this question, we have generated a series of ORN populations with increasing dissimilarity of the ORN within populations and compared its mutual information.

Before we analyze the information transmission, there are two issues that require special attention, namely, the stimuli distribution and the quantification level of the frequency axis. To the best of our knowledge, a statistical distribution for the input odor concentration to the ORNs has not been studied. Therefore, our position is to provide the simpler stimuli distribution to our populations: a uniform distribution. About the second issue, the discretization of the frequency axis, we are in a similar situation. Due to a lack of experimental studies, as far as we know, about the frequency resolution of ORNs, we have chosen to split the frequency axis in 10 bins. It is worth noticing that any other selection would only scale our results on mutual information.

To evaluate the mutual information of these populations, we generate random stimuli (concentration values) from a uniform distribution between -3 and -7 log units (dose C). The mutual information is computed over the accumulated response of all ORNs to all stimuli (Alkasab et al. 2002). To isolate the contribution of each one of the parameters F_m, n, and C_T to the mutual information, we generated three series of ORN populations in which only one of the parameters is varied and the other two remain constant.

It is convenient to study the effect of each one of the parameters individually to determine its contribution to the information transmission. Figure 2.2a shows the relationship between spiking frequency and odor concentration as expressed by Equation 2.1. This relationship depends upon three parameters: threshold concentration (C_T), n factor (n), and maximum frequency (F_M). F_M and C_T are described in Figure 2.2a, and the n factor determines how fast the curve reaches the maximum frequency (F_M). The different firing frequency characteristics of different ORNs in the olfactory epithelium can be described choosing different values of these three parameters. Therefore, we can generate a population of ORNs by assigning certain statistical distributions to each one of the parameters.

2.2.1.1 Variation of the n Factor

In this first series of populations, the ORN within the population will have the same $F_m = 2.44$ Hz and $C_T = -6$ log unit, but the n parameter will be generated from a lognormal distribution with $\mu = 2.44$ log units (these experimental values are extracted from Rospars et al. (2003)). The standard deviation (σ) of this distribution is used as a parameter to control the similarity of the ORNs. For $\sigma = 0$ log units, all the ORNs have the same parameters, so we have a homogeneous population. As σ increases, the variety of ORNs will increase accordingly. We have generated a series of populations for increasing values of σ. Each population comprises 10,000 neurons and is excited with 1000 stimuli. Figure 2.3a shows the mutual information of these populations in terms of their standard deviation. The mutual information increases initially until it reaches a maximum around $\sigma = 1$ log unit; after that, it decreases monotonically.

A first conclusion we can extract from this result is that a homogeneous population ($\sigma = 0$ log units) provides less efficient encoding than other populations with some variety of the n factor. This partially answers the question posed at the beginning of this section. A second conclusion that can be extracted is that there exists a

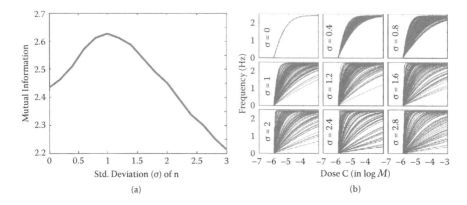

FIGURE 2.3 (a) Mutual information of the static populations with different values of the variance of the ORN n parameter. (b) Frequency-dose curves for populations with increasing value of σ.

σ value that provides an optimum encoding of the concentration information. This result can be further analyzed by observing the concentration-frequency curves for the different populations. Figure 2.3b shows these curves for several values of σ. For $\sigma = 0$ log units we have that all the ORNs share the same curve. As σ increases, the population provides a better coverage of the input-output space, making more homogeneous the output distribution, which in turn increases the mutual information. The population reaches a point ($\sigma > 1$) where the number of ORNs with very high or very low n is high enough to worsen the homogeneity of the output distribution.

2.2.1.2 Variation of the Threshold Concentration C_T

In this second series of populations, the values of n and F_m are fixed to 0.5 and 2.44 Hz, respectively, and C_T is generated from a uniform distribution. The range of the uniform distribution is in this case used to control the similarity of the ORNs. As in the previous subsection, the null range generates a homogeneous population, and increasing its value, the ORN similarity is decreased. Twenty-one populations were generated for range values between 0 and 4 log units. Each population comprises 1000 neurons and is excited with 1000 stimuli. Figure 2.4a shows the mutual information for populations with increasing width of C_T. The mutual information has a behavior similar to that in the previous subsection, reaching a maximum for range = 1.6 log units.

Similarly to the previous subsection, two conclusions can be extracted from this result. First, having a nonhomogeneous population represents an advantage for the concentration information coding. Second, there exists an optimum width value for the concentration information coding. Figure 2.4b shows the frequency-dose curves for the different populations generated. We can see how as the range increases, the population covers more frequency-dose space and provides a more uniform distribution at the frequency space, which in turn increases the mutual information. However, as the new ORNs have increasing C_T, they cover less dose range, and this produces a decrease of the homogeneity at the output space. These two factors compete with each other, producing the maximum of the mutual information.

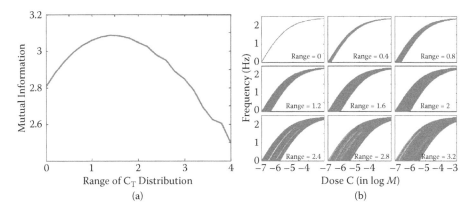

FIGURE 2.4   (a) Mutual information of the static populations with different values of the C_T distribution range. (b) Frequency-dose curves for populations with increasing value of the C_T range.

2.2.1.3   Variation of the Maximum Frequency F_m

Finally, we generated a series of populations fixing the $n = 0.5$ and $C_T = -6$ log units parameters and generating F_m from a lognormal distribution with $\mu = 2.44$ Hz. The standard deviation of the distribution is again used to control the similarity of the ORN within a population. Sixteen populations are generated with standard deviation between 0 and 3. Each population comprises 1000 neurons and is excited with 1000 stimuli. Figure 2.5 shows the mutual information of these populations. In this case, the mutual information does not have a maximum and decreases monotonically. This result denotes that in terms of the maximum frequency parameter, a homogeneous population is more advantageous than a nonhomogeneous population of ORN.

Summing up, this partial theoretical study of the ORN populations predicts that nonhomogenous populations of ORNs encode more efficiently odor concentration information than homogeneous ones. Furthermore, this analysis provides theoretical

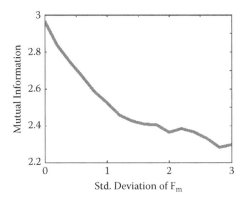

FIGURE 2.5   Mutual information of static population with increasing values of the standard deviation of F_M.

evidence for the existence of an optimum spread of the ORN parameters to obtain an efficient odor information encoding. Interestingly enough, these results can serve to explain why the olfactory system has a nonhomogeneous population of ORNs to measure odor concentration.

2.2.2 DYNAMIC ORN POPULATIONS

To complete the previous study, we generated a dynamic ORN population based on an ORN model that considered also time-dependent parameters (latency and duration). It is worth noticing that this dynamic model is considerably different from the static one, and therefore required a different computation of the coding efficiency. The activity of these populations is expressed in timescale since each ORN has its spiking train perfectly defined in terms of latency, duration, and frequency. Figure 2.6 shows the spike trains generated by a population of 700 ORNs, where frequency, latency, and spike duration of each ORN are randomly generated from the distributions described in Rospars et al. (2003). Each row is the spike train generated by one ORN along the time where spikes are represented by blue dots.

To measure the odor concentration information encoded by the ORN populations, we have followed the subsequent procedure explained in Figure 2.7. We have generated populations of 5000 ORNs by assigning to each ORN a different combination of the three parameters mentioned before: spiking frequency, latency, and duration. The values of these parameters are randomly selected from their statistical distributions found by Rospars et al. (2003). To study these populations, it would be desirable to excite them with the distribution of concentrations that is presented to the biological ORN population. However, since there are no studies in this respect, to the best of our knowledge, we have used a random log-uniform distribution in the concentration range of [–7, –3] to excite the ORN populations. Then, the spike trains obtained from each ORN for each concentration are used to compute a histogram in the time axis. The histograms are computed using 10 ms as a bin width. We compute the mutual information of the population from the histogram obtained.

The generation of the ORN population has been driven by the objective of evaluating how efficiently the population encodes for odor concentration information. We would like to evaluate this coding efficiency as the population becomes more

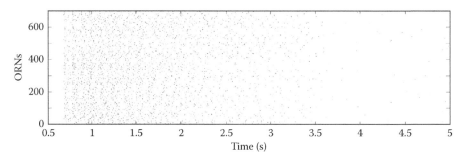

FIGURE 2.6 Raster plot of the spike trains produced by a population of 700 ORNs. Each of the rows represents different ORNs, and spikes are represented by dots.

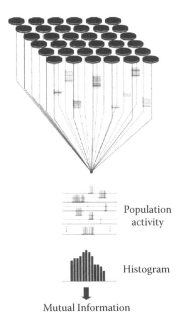

Population
activity

Histogram

Mutual Information

FIGURE 2.7 Computation of the mutual information of an ORN population. From the reduced ORN model we built ORN populations and obtained their time responses. The mutual information is then computed from the histogram of the ORN population response.

heterogeneous. To generate populations of increasing levels of heterogeneity, the parameters of all ORNs are fixed, except for one that is allowed to be different across ORNs. The range of values that this changing parameter is used to increase the heterogeneity of the ORN populations. We have generated a series of populations in which the first one has a null parameter range, meaning that all ORNs are equal. The subsequent populations are obtained by increasing the range of the parameters, therefore increasing their heterogeneity.

We generated two series of populations varying the parameters n and C_T of Equations 2.1 to 2.3. Notice that the first parameter only affects the firing frequency (Equation 2.1), not latency and duration. We have chosen this parameter in order to compare the behavior of this complete model with that of the static model, which only considered firing frequency. The second parameter used to generate a series of populations (C_T) allowed us to evaluate the behavior of the entire model since firing frequency, latency, and duration depend on C_T.

Figure 2.8 shows the mutual information computed on the two series of populations as the range of n and C_T is increased. In both cases the mutual information reaches a maximum for a certain value of the range. However, it is important to notice that the amount of bits increased from the null range to the maximum is higher when we vary C_T. These results are consistent with those obtained with the static ORN model.

Several conclusions can be drawn from these results. The main of these conclusions is that the information transmitted by the ORN population is maximized when

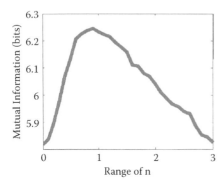
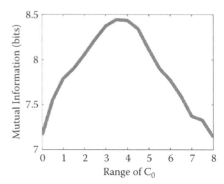

FIGURE 2.8 Mutual information of two series of dynamic ORN populations as the dissimilarity of the populations increases.

the population is heterogeneous (homogeneous population correspond to zero range value). This is theoretical evidence that supports the superior odor information coding efficiency of heterogeneous ORN populations. Second, there exists an optimal value of dissimilarity within ORN populations that maximizes information transmission. This dissimilarity level may be the one that is found in biological ORN. Finally, a last conclusion can be drawn comparing these results with those obtained with the static model. Using the static model, the increase of mutual information between the homogeneous populations and the optimum heterogeneous one was 0.2 bit, whereas in the current results the increase goes up to about 1.5 bits. This gives an idea of the larger amount of information that is accounted for in the dynamic model. Therefore, the time-dependent parameters that we have added to the ORN model are able to encode for additional information that was not captured by the static model.

2.3 STUDY OF ODOR INTENSITY CODING IN THE GLOMERULI

In this section we continue the coding efficiency study of the olfactory pathway extending the information theoretic analysis of the ORN population with the next stage of the olfactory pathway. This stage involves a projection of ORN axons to spherical regions of the neuropil called glomerulus. This projection is highly ordered since it has been observed that ORNs express the same olfactory receptor project to the same glomerulus. This allows studying odor intensity by focusing only on one glomerulus. Within these spherical structures we can find not only ORN axons, but also dendrites of mitral/tufted cells and periglomerular cells. The interaction of the existent types of cells defines a complex behavior that is not totally understood at this date. However, it is generally considered that there exists an inhibitory effect between ORN axons mediated by either periglomerular cells or microcircuits formed by ORN axons and mitral cell dendrites. These can be modeled as a lateral interaction between the incoming ORN axons before transmitting the odorant information (Figure 2.9). We used this simple interaction model to study the odor concentration information transmission, performing experiments first in a reduced model and second in a more comprehensive one.

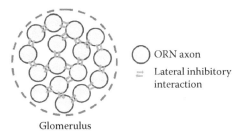

ORN axon

Lateral inhibitory interaction

Glomerulus

FIGURE 2.9 Glomerulus. The glomerulus is a neuropil where the ORN axons project. There exist inhibitory interactions between ORN axons in the glomerulus.

2.3.1 REDUCED ORN POPULATION

To gain some insight on the effect of the lateral inhibition in the information encoding, we first studied a reduced population to facilitate the interpretation of the results. This toy population is formed by 13 ORNs characterized by their dose-response curves. Figure 2.10a shows the response curves that have been obtained from Rospars et al.'s (2003) model considering only the frequency response. The parameters used to generate the ORNs were the following: constant values of $F_m = 10$ Hz and $C_T = -7$, and variable values of n from 0.5 to 3 with a step of 0.2.

The effect of the lateral inhibition is obtained by subtracting pairwise the dose-response curve of those ORNs that are connected. This is a first and simple approach to model these interactions, but consistent with the level of abstraction of our population model. The connectivity in our toy problem has been set to link all ORNs to that of a lower frequency response. Figure 2.10b shows the frequency response curve of the pairwise linked ORNs.

To compute the mutual information (MI) of the ORN population before and after the inhibitory stage, we present the population with a stimulus of random

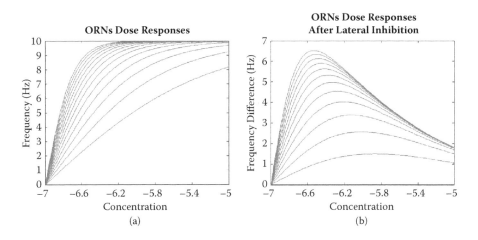

FIGURE 2.10 Dose-response curves of toy population before (a) and after (b) lateral inhibition.

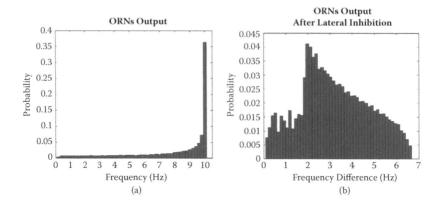

FIGURE 2.11 Frequency probability of the toy population output before (a) and after (b) inhibition for a uniformly randomly distributed [−7, −3] input.

concentrations extracted from a uniform distribution in the range of [−7, −5]. The population response to these stimuli is shown in Figure 2.11 as the probability of occurrence of each frequency for the ORN population before (a) and after (b) inhibition. We can see how the effect of the inhibition distributes more evenly the probability of the frequencies. The probability distribution before inhibition (Figure 2.11a) is clearly peaked at 10 Hz because of the saturation of all dose-response curves at 10 Hz (Figure 2.10a), whereas the frequency probability after inhibition (Figure 2.11b) is more evenly distributed. In terms of information transmitted (MI), this translates into a larger amount of information transmitted after the lateral inhibitory layer. The MI values of the ORN population before and after inhibition are 4.35 and 5.46 bits, respectively. The information transmitted has increased by 1.1 bits due to the lateral inhibition.

2.3.2 Comprehensive ORN Population

To make a more thorough study of the effect of the lateral inhibition on the information transmission, we have built a population of ORNs that includes all the variability of ORNs experimentally observed. We have modeled a population of ORNs that project to a single glomerulus and analyzed the concentration information transmitted before and after the inhibitory interaction at the glomerulus. The ORN model used is that proposed by Rospars et al. (2003) for the firing frequency (Equations 2.4 to 2.6). In this mathematical model there are three parameters (F_m, C_T, n) that determine the behavior of the individual ORNs. The statistical distributions of these parameters in the frog olfactory epithelium were experimentally found also in this paper. The maximum frequencies (F_m) are randomly extracted, for each ORN of the population, from a lognormal distribution with mean of 2.44 Hz and a standard deviation of 0.44 Hz. The C_T parameter is randomly selected from a uniform distribution in the range of −7 to −3. Finally, n follows the inverse of a lognormal distribution centered at 0.2 and with a 1.16 standard deviation. We have used these statistical

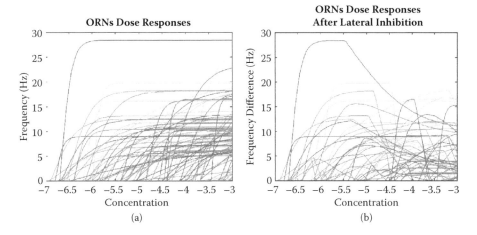

FIGURE 2.12 Dose-response curves of complete population before (a) and after (b) lateral inhibition.

distributions to generate a population of 20,000 ORNs that project to the same glomerulus. Figure 2.12a shows the dose-response curve of 100 of these 20,000 ORNs. Before we explain the model of inhibition we have used, it is important to remark that our ORNs are characterized by its dose-response curve, which is the spiking frequency in terms of the concentration of the presented odor.

To model the effect of the inhibitory stage, we randomly connected ORN axons limiting the number of connected axons to one. Each axon will be inhibited by another axon, and only by that one. A difficult point was to model the effect of the inhibitory interactions in the dose-response curves. Dose-response curves assume constant frequency of the ORNs' response for certain odor concentrations, and the interaction of two ORNs with different frequencies may lead to responses with nonconstant frequency responses. However, taking into consideration the time integrating effect of the neurons, we can consider that the result of an inhibitory incoming spike train is to reduce the spike frequency of the target neuron proportionally to the frequency of the inhibitory neuron. Therefore, the dose-response curve after inhibition can be obtained as a direct subtraction of the dose-response curves of the target and inhibitory ORNs. Figure 2.12b shows the dose-response curves of the ORNs after being inhibited by another ORN. In this figure only 100 of the 20,000 ORNs are shown to facilitate visualization.

At this point we can determine the odor concentration information transmitted after the lateral inhibitory stage computing the mutual information. To do so, we have presented the ORN population with 100 different odor concentrations randomly extracted from a uniform probability distribution in the range of [–3, –7]. After computing the dose-response curves of all ORNs of the population, we inhibit the ORN axons following the lateral connections, obtaining the final dose-response curves. The output frequency probability is obtained concatenating the response of all ORNs and computing the histogram from 0 to 40 Hz in 50 bins and normalizing afterwards.

We have studied the effect of the lateral inhibition using a parameter to regulate the intensity of the inhibitory interaction. This parameter will take values between 0 (no inhibition) and 1 (maximum inhibition). The inhibition is maximum when the dose-response curve of the inhibitory neuron is directly subtracted from the dose-response curve of the target neuron. We computed the response of our ORN population for 11 values of the inhibition intensity starting from 0, with an increasing step of 0.1, until 1 is reached. Figure 2.13 shows the output frequency distributions of the ORN populations for the increasing values of inhibition. The first subfigure (top left) represents the frequency distribution as if no inhibition was applied. As the inhibition intensity increases, the central peak of the figure decreases, vanishing almost totally for the maximum inhibition intensity. So the effect of the lateral inhibition is to reduce the importance that the frequencies around 10 Hz have at the output of the population and redistribute this activity among the other frequencies.

Computing the mutual information of the distributions of Figure 2.13, we obtain the results of Figure 2.14. This figure shows the mutual information in terms of the intensity of the inhibitory interaction. A maximum of the mutual information is obtained for 0.3. Looking at Figure 2.13, this value corresponds to a frequency distribution where the main peak is almost leveled, with the smaller peak forming a broader peak that distributes the output more evenly among the different frequencies. This shows that a certain amount of inhibition is able to redistribute the output frequencies more evenly, and therefore enhance the amount of odor concentration information that is transmitted. It is important to notice that the increase of mutual information found is not very high, and it does not seem to account completely for the positive effect that lateral inhibition is thought to have in the processing of the odorant information. However, the present results are evidence that lateral inhibition can increase the odor intensity information transmitted by the early stages of the olfactory system, and more significant increases of the mutual information may be obtained by using more accurate models of the inhibitory interaction. The validity of this conclusion is limited by the simplicity of the ORN and inhibitory model adopted in these simulations. A more detailed model of ORN and lateral inhibition may reveal some information transmission advantage not captured here.

2.4 STUDY OF ODOR INTENSITY/IDENTITY CODING IN AN ORN POPULATION

As mentioned in Section 2.1.1, odor identity is captured by the olfactory system using a combinatorial coding strategy. Individual ORNs are able to bind to a number of ligands with different affinities, and any ligand in turn is bound by a number of ORNs. Therefore, each odor (specific collection of ligands) elicits a particular pattern of ORN activation across the ORN population (Buck and Axel 1991). The objective of this section is to evaluate the advantages in information transmission of such a coding strategy to encode for odor identity and intensity. In the previous sections we have limited our studies to odor intensity by using an ORN model that considered the sensitivity of the ORNs to certain odors, but not the selectivity to different odors. In this section we will use an ORN model that takes into account both. This model was proposed by Rospars et al. (2003) and is described by Equations 2.4 to 2.6.

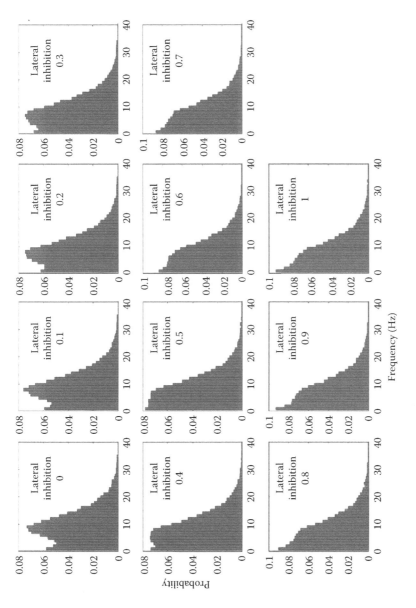

FIGURE 2.13 Output frequency distributions for different lateral inhibition intensities.

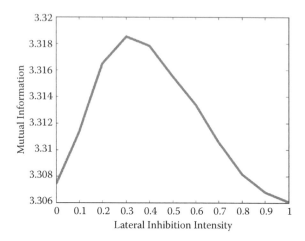

FIGURE 2.14 Mutual information of an ORN population in terms of the intensity of the lateral inhibition within the glomerulus.

2.4.1 ORN AFFINITY DISTRIBUTION

To build ORN populations using the previous ORN model, we are still missing a piece of the puzzle. We need to know the distribution of binding affinities of the ORNs to certain ligands within an ORN population. Lancet et al. (1993) proposed to determine this distribution of affinities toward certain ligands using a combinatorial approach. They modeled receptors and ligands as having B binding subsites where each one could be in one of S possible configurations. They assumed a one-to-one correspondence between the B subsites of the receptor and those of the ligand. The binding was successful if at least L of the B subsites of ligand and receptor where in a complementary configuration. Assume that all receptors and ligands have the same B and S, and also that all subsite configurations are equiprobable. This model represents the receptor-ligand binding as composed by B independent and identical elementary experiments where the number of successful interactions (bindings) measures the affinity of receptor and ligand. Therefore, the probability distribution of ORNs for certain fixed ligands in terms of the number of subsites bound (L, affinity) follows a binomial distribution.

This affinity distribution allows modeling the interaction of a receptor with any ligand. In this way, we can characterize each ORN of a population with a vector of N affinities that describes the interaction between the ORN and N different ligands. To build an ORN population that is in accordance to this distribution, we generate groups of N values (affinity vectors) randomly extracted from a continuous binomial distribution. The affinity vector characterizes the frequency response of the ORNs, since the affinity parameter corresponds to the parameter Mg/2 in Equation 2.4. Figure 2.15b shows the dose-response curves of an ORN to the exposure of four ligands. These curves are characterized by an affinity vector of four values. This ORN has higher affinity to odor C than the rest. The following odors, in order

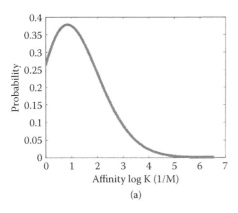
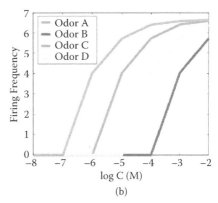

(a)

(b)

FIGURE 2.15 ORN affinity distribution. (a) Distribution of affinities given by a continuous binomial distribution. (b) Dose-response curves of an ORN characterized to respond to four odors.

of increasing affinities, are A, D, and finally B, which has the lowest affinity to this ORN. A ligand with higher affinity is able to elicit activity (firing frequency) of the ORN, starting from lower concentrations, and produces higher firing rates for the same amount of concentration than other ligands, as we can see for ligand C. To determine the response of the ORN of a mixture of odors, we assume hypoadditivity and consider that the response of the ORN is the highest response to any of the mixture components.

2.4.2 ORN SPECIFICITY STUDY

With this more complete population model, we study again the effect of the specificity of ORN in the coding efficiency. We try to answer the same question that is in previous sections: How does the specificity of receptors in an ORN population affect the quality and concentration of information transmitted by the population? To address this question, we generated a series of ORN populations with varying specificities of the ORNs within populations and compared their mutual information.

The regulation of the specificity of the ORNs is mediated by a modulation factor p that multiplies the ORN affinity vectors. This modulation factor is obtained with Equation 2.8:

$$p_m = \left(\frac{1}{m}\right)^{10^{-t}} \quad m = 1, 2, 3, ..., 10; \quad t = -1, -0.8, -0.6, ..., 0.6, 0.8, 1 \quad (2.8)$$

where m is the ligand ranging from 1 to 10, and t is a parameter that modulates the specificity of the ORN. It ranges from -1 (high specificity) to 1 (no modulation). Figure 2.16a shows the shape of p for different values of t. When $t = -1$ the value of p is zero or has a low value for most of the ligands, except for ligand 10, which is 1. That means that this ORN will have a reduced response to any ligand except ligand

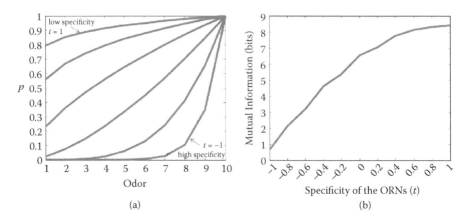

(a)   (b)

FIGURE 2.16   Specificity study. (a) Modulation factor (p) for six values of t. High specificity for $t = -1$ and low specificity for $t = 1$. (b) Mutual information of the populations as the ORN specificity decreases.

10, becoming very specific to detect it. As the value of t increases, so increases the number of ligands and the contribution of all ligands in p, resulting in a straight line for $t = 0$. This makes the ORN response less specific. When the value of t reaches 1, all ligands contribute almost equally, being equal or close to one value. This results in no modulation of the original ORN affinity vectors.

We generated populations of varying specificity composed of 100 ORNs able to respond to 10 ligands. The response of each ORN was determined by randomly generating 10-dimensional affinity vectors from a continuous binomial distribution (Figure 2.15a). Once these ORN affinity vectors are defined, we modulate them with the p factor (Equation 2.8) to regulate the specificity of the ORNs within each population. We built 11 different populations taking values of the t parameter from -1 (high specificity) to 1 (no modulation), separated by 0.2 each. We presented the population with 100,000 randomly generated odors (mixtures of ligands), where each odor is defined as a 10-dimensional vector with the concentration of each component. The concentrations of the ligands are randomly generated following a uniform distribution within the range of -3 to -7 log units.

To determine the efficiency of these populations when encoding for odor quality and odor intensity with these stimuli, we computed the mutual information between the input and output signals as described in Section 2.1.3. Figure 2.16b shows the results obtained for the 11 populations. The mutual information for the population with the more specific ORNs ($t = -1$) is the lowest. The mutual information increases then as the specificity of the ORNs decreased within the populations, and it saturates for a maximum value when t reaches 1. This result clearly shows that populations with specific ORNs encode less information than those with less specific ORNs. The saturation of the mutual information value for t approaching 1 is due to the absence of modulation from p. It is close to 1 for all ligands. So, the population exhibits the amount of specificity generated when the affinities are randomly chosen from a binomial distribution. This reflects that this distribution is optimal for the transmission of information.

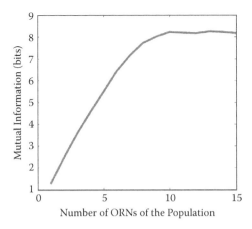

FIGURE 2.17   Population dimensionality study. The mutual information to encode for 10 odors as the dimensionality of the population increases from 1 to 15.

2.4.3   POPULATION DIMENSIONALITY STUDY

The second study we performed is motivated by the following question: How does the dimensionality of the ORN population affect the information transmission? To address this question, we generated a series of populations with an increasing number of ORNs, from 1 to 15. Each ORN was able to respond to 10 different ligands, and the affinity vectors characterizing this response were obtained with a continuous binomial distribution. In order to compute the input-output information transmission, we present the model with 10,000 input mixture odors. These odors are generated, as in the previous section, as a mixture of the 10 ligands that are randomly extracted from a uniform distribution within the range of –3 to –7 log units. Figure 2.17 shows the value of the input-output mutual information for the population for an increasing number of ORNs. The mutual information increases with the number of ORNs reaching saturation for 10 ORNs. This result shows that beyond 10 neurons, the information encoded by the population does not increase further.

From these results we can draw several conclusions. The first is in line with those obtained in the studies of odor intensity. It is much more efficient to encode the information of a mixture of odors with nonspecific receptors. This analysis provides theoretical evidence for the existence of an optimum spread of the ORN parameters to obtain an efficient odor information encoding. From the second study we can conclude that the dimensionality of the population determines the coding capacity of the population.

2.5   CONCLUSIONS

From the three information theoretic studies presented in this chapter, we can draw several conclusions. First, this work presents theoretical evidence that supports the advantages of nonspecific ORN populations for the encoding of odor intensity as well as for odor identity. Furthermore, it has been shown that there exists a value of the variability of the ORNs that maximizes the transmission of odor information. In the case of odor intensity, this has been shown for static and dynamic models of ORNs.

Second, the lateral inhibition at the glomerular level has been shown to improve the coding efficiency of odor intensity. The limited improvement observed in this work is probably explained by the simplicity of the inhibition models. More detailed models of lateral inhibition can lead to a more significant improvement on information coding efficiency.

Finally, it is worth noting that the studies conducted in this chapter did not consider an important aspect of neural activity: neural noise in the signals. Considering noise, it would be possible to provide a more realistic estimate of the amount of information transmitted by the early olfactory pathway. We can consider the present results as an upper bound of this information.

REFERENCES

Alkasab, T. K., J. White, and J. S. Kauer. 2002. A computational system for simulating and analyzing arrays of biological and artificial chemical sensors. *Chem. Senses* 27(3):261–275.

Aungst, J. L., P. M. Heyward, A. C. Puche, S. V. Karnup, A. Hayar, G. Szabo, and M. T. Shipley. 2003. Centre-surround inhibition among olfactory bulb glomeruli. *Nature* 426:623–629.

Axel, R. 1995. The molecular logic of smell. *Sci. Am.* 273:154–159.

Buck, L., and R. Axel. 1991. A novel multigene family may encode odorant receptors: A molecular basis for odor recognition. *Cell* 65:175–187.

Cover, T. M., and J. A. Thomas. 2006. *Elements of information theory.* New York: John Wiley & Sons.

Doty, R. 1991. The olfactory system. In *Smell and taste in health and disease*, 175–203. New York: Raven Press.

Grosmaitre, X., A. Vassalli, P. Mombaerts, G. M. Shepherd, and M. Ma. 2006. Odorant responses of olfactory sensory neurons expressing the odorant receptor MOR23: A pacth clamp analysis in gene-targeted mice. *Proc. Natl. Acad. Sci. U.S.A.* 103:1970–1975.

Hildebrand, J. G., and G. M. Shepherd. 1997. Mechanisms of olfactory discrimination: Converging evidence for common principles across phyla. *Annu. Rev. Neurosci.* 20:595–631.

Lancet, D., E. Sadovsky, and E. Seidemann. 1993. Probability model for molecular recognition in biological receptor repertoires: Significance to the olfactory system. *Proc. Natl. Acad. Sci. U.S.A.* 90:3715–3719.

Ma, M., and G. M. Shepherd 2000. Functional mosaic organization of mouse olfactory receptor neurons. *Proc. Natl. Acad. Sci. U.S.A.* 97(93):12869–12874.

Ressler, K., S. Sullivan, and L. Buck. 1994. Information coding in the olfactory system: Evidence for a stereotyped and highly organized epitope map in the olfactory bulb. *Cell* 79:1245–1255.

Rieke, F., D. Warland, R. Steveninck, and W. Bialek. 1997. *Spikes: Exploring the neural code.* Cambridge, MA: MIT Press.

Rospars, J. P., P. Lánský, A. Duchamp, and P. Duchamp-Viret. 2003. Relation between stimulus and response in frog olfactory receptor neurons in vivo. *European Journal of Neuroscience* 18:1135–1154.

Shannon, C. E., and W. Weaver. 1949. *The mathematical theory of communications.* Urbana: University of Illinois Press.

Shepherd, G., and C. A. Greer. 2004. Olfactory bulb. In *The synaptic organization of the brain*, ed. G. Shepherd, 159–204. Oxford: Oxford University Press.

Shepherd, G. M. 1987. A molecular vocabulary for olfaction. *Annals of the New York Academy of Sciences* 510:98–103.

Shepherd, G. M. 1994. Discrimination of molecular signals by the olfactory receptor. *Neuron* 13:771–790.

Vassar, R., S. K. Chao, R. Sitcheran, J. M. Nuñez, L. B. Vosshall, and R. Axel. 1994. Topographic organization of sensory projections to the olfactory bulb. *Cell* 79:981–991.

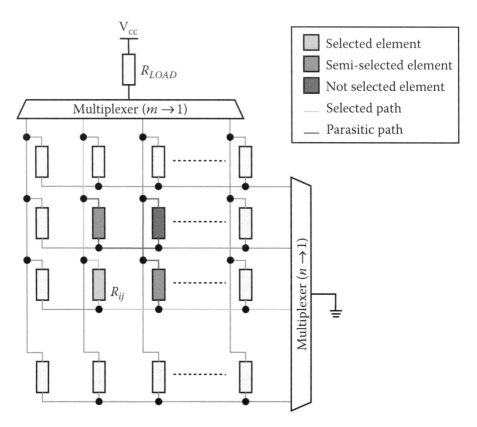

COLOR FIGURE 3.3 The matrix configuration adopted for the sensor array reduces the number of electrical wires needed for the sensor connections. The readout circuit is more complex and powers the sensors, while compensating for the electrical cross talk. For element R_{ij}, the column multiplexer (top), selects the jth column to provide power through R_{LOAD}. Simultaneously, the row multiplexer selects the ith row and connects it to the ground, so that the green path is selected. All the elements of the jth column are powered, while all the elements of the ith row are connected to the ground. There are semiselected tracks (orange) and nonselected paths (red) that result in unwanted resistances connected in parallel to R_{ij}.

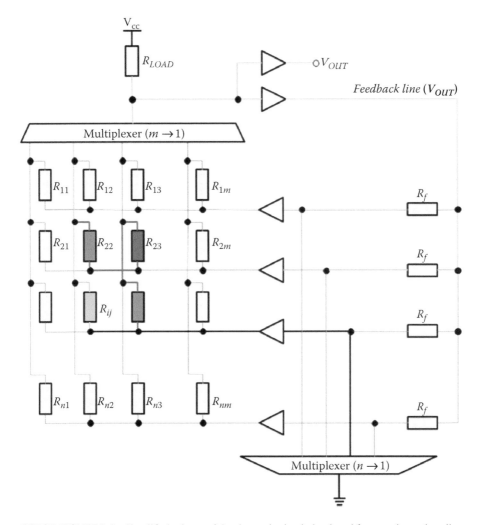

COLOR FIGURE 3.4 Simplified scheme of the electronic circuit developed for scanning and reading the large sensor array. In order to read the element R_{ij}, the column multiplexer selects the jth column and connects it to the power through the load resistance. The row multiplexer connects the ith row to ground, while all the other nonselected rows are connected to the output voltage by means of feedback buffers. Hence, the selected sensor is correctly powered, while no current can pass through the nonselected resistances, because of the zero-voltage difference across them. This is illustrated in the figure, where the semiselected element R_{22} is at V_{out} in its first terminal through the column multiplexer and virtually at the same potential in the second terminal because of the feedback line. In this way no current can flow through that element, disconnecting the entire parasitic path from the element being read.

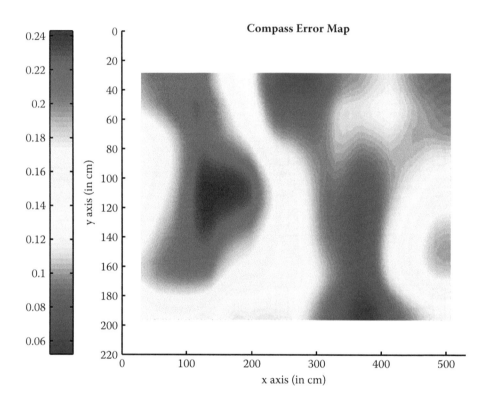

COLOR FIGURE 4.22 Heat plot illustrating the deviation of the compass readings in comparison to the simulated compass. The areas of high deviation are marked in red, and of low in blue.

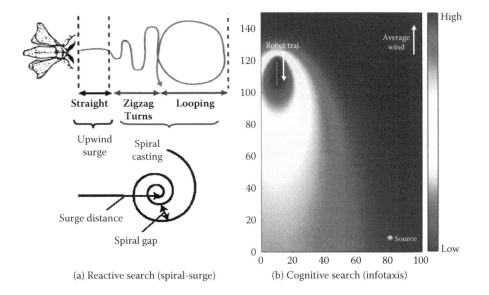

(a) Reactive search (spiral-surge)   (b) Cognitive search (infotaxis)

COLOR FIGURE 5.2 Reactive and cognitive search strategies. (a) Spiral-surge (reactive search) is inspired by the odor-modulated anemotactic response of male moths in pheromone plumes. Top: A typical trajectory of a silkworm moth consists of two phases, an upwind surge in the presence of the odor and a casting phase in its absence. During casting, the moth performs first zigzag turns across the wind in the hope of regaining contact with the odor, and then switches to circular motion patterns. (Reprinted with modification from Kanzaki, R., Ando, A., Sakurai, T., and Kazawa, T., *Adv. Robotics,* 22(15), 1605–1628, 2008. With permission.) Bottom: In spiral-surge, the casting phase is modeled as an Archimedean spiral with a gap parameter (Hayes et al. 2002; Lochmatter et al. 2008). The surge phase requires knowing the wind direction and has a single parameter, the surge distance. (b) Infotaxis (cognitive search). Example of belief (probability map) for the location of the source after 10 steps (red dots). No cues are detected in that time. Locations in front of the agent become less probable as the agent navigates forward without encounters, thereby increasing the likelihood of locations on the sides. The shape of the belief (Gaussian-like) is inferred from the physical description of how cues spread in the environment when transported by the wind.

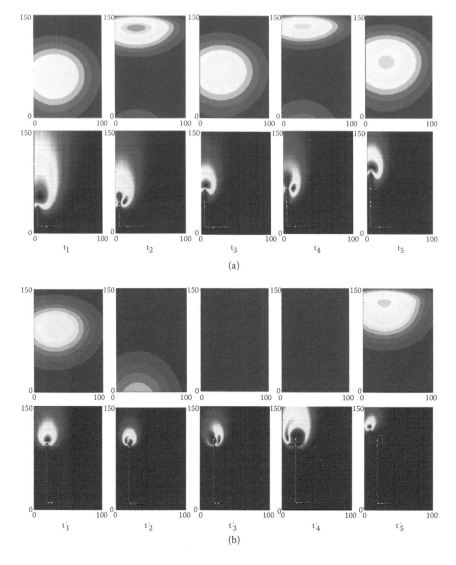

COLOR FIGURE 5.4 Navigation patterns observed when infotaxis is confronted to a pulsed source (fast pulses in a, slow pulses in b). Top rows represent snapshots of the simulated environment at different times (pulsed source located in $r_0 = (25, 2)$, wind blows downward), and bottom rows are for the corresponding source distribution (belief function). False blue and red colors correspond to low and high probabilities, respectively. The path of the robot is superimposed to the map as consecutive red dots when there is no detection, and green dots for detections. (a) High-frequency pulsed patches provoke new detections before the robot starts spiraling. High probabilities are frequently updated and assigned to upwind locations (at times t_1, t_3, and t_5), hence pushing the agent forward. (b) Between pulses, long intervals of clean air—during which no detections arise—compel the agent to explore regions where previous detections were recorded. Probabilities updates take the form of concentric ellipses that spread as the robot navigates around them, as clearly seen at times t_2', t_3', and t_4'. (Note that such behavior—although with much smaller radius—may also be recorded for the fast-pulsed case when the agent is close to the source, due to an excessive amount of detections that push him to switch to exploitation mode.) (Reprinted from Moraud, E. M., and Martinez, D., *Front. Neurorobotics*, 4: 1, 2010, DOI: 10.3389/fnbot.2010.00001.)

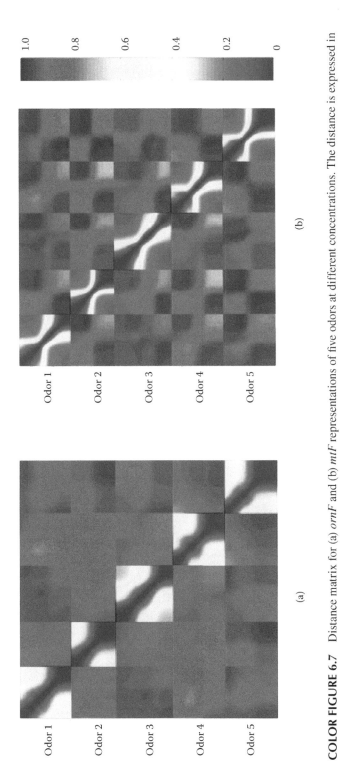

COLOR FIGURE 6.7 Distance matrix for (a) *ornF* and (b) *mtF* representations of five odors at different concentrations. The distance is expressed in the interval [0, 1] as a quantity complementing the correlation coefficient to unity.

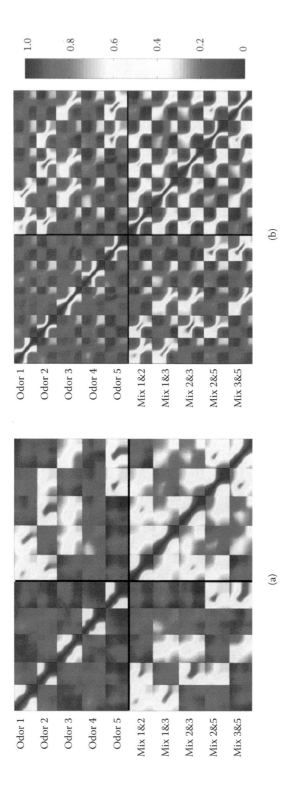

COLOR FIGURE 6.8 Correlation-based distance matrices for (a) *ornF* and (b) *mtF* representations of five single components and their five binary mixtures (1&2, 1&3, 2&3, 2&5, and 3&5). The horizontal and vertical black solid lines separate the areas corresponding to single odor and binary mixtures.

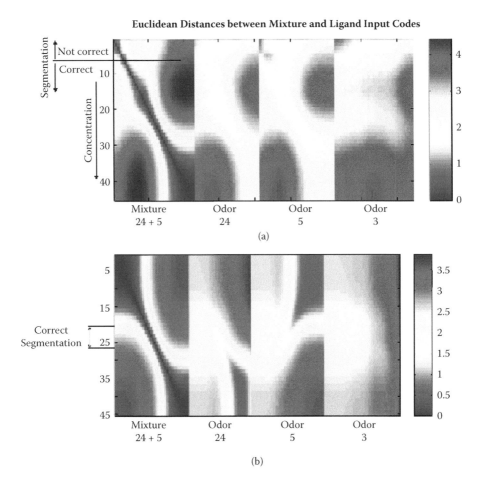

COLOR FIGURE 6.14 Euclidean distances between representations of a mixture of two odors, its components, and a single-ligand odor that is not part of the mixture at various concentrations for two different mixture schemes: (a) constant concentration ratio and (b) complementary proportion mixing. Also shown is the segmentation result. Segmentation is successful when the Euclidean distances between odor components and the mixtue are low at the same time as distances are high between the mixture and odors not present in the mixture.

3 Mimicking Biological Olfaction with Very Large Chemical Arrays

Mara Bernabei, Romeo Beccherelli, Emiliano Zampetti, Simone Pantalei, and Krishna C. Persaud

CONTENTS

ABSTRACT

Olfactory receptor neurons in the mammalian olfactory system transduce the odor information into electrical signals and project these into the olfactory bulb. In the biological system the huge redundancy and the massive convergence of the olfactory receptor neurons to the olfactory bulb are thought to enhance the sensitivity and selectivity of the system. To explore this concept, a modular polymeric chemical sensor array consisting of up to 65,536 sensors belonging to tens of different types was built. To interface such a large sensor array, a topological array configuration with n rows and m columns was adopted in order to reduce the number of wiring connections, and 16 subarray modules are simultaneously read to make the acquisition speed comparable to biological olfaction. The matrix array configuration is not without consequence. It introduces electrical cross-talk which makes the response of each sensor dependent also on the responses of other sensors in the array. An analysis of the electrical cross-talk and the solution adopted to compensate for it are here presented together with some discussions on advantages and limitations of the scanning algorithm. Preliminary experiments were carried out on the prototype to test its capability in terms of odor discrimination, classification, and binary mixtures segmentation. The data were analyzed using multivariate methods and the results indicate that the large scale array developed can emulate several aspects found in biological olfaction.

3.1   INTRODUCTION

In the 1980s work began on a new instrument, mimicking some concepts of the mammalian olfactory system (Persaud and Dodd 1982). Since then, these types of devices have been investigated in depth and developed for many applications, such as in the food and pharmaceutical industries, in environmental control, clinical diagnostics, and more (Pearce et al. 2003). The rationale behind the functioning of these instruments has to be found in the use of a few different but relatively nonselective chemical sensors. Their responses are not directly correlated to a specific chemical substance, but rather to the whole repertoire of chemical information contained in an odor, which is often a mixture of many substances. This somewhat resembles the way the biological olfactory system deals with odorants. However, biological olfaction outperforms current chemical analysis instrumentation in specificity, response time, detection limits, coding capacity, stability over time, robustness, size, power consumption, and portability. This is mainly due to the unique architecture of the olfactory pathway, characterized by a high level of sensor redundancy, early combinatorial coding, and exceptional information processing mechanisms.

The fabrication of instrumentation having similar performance and efficiency to those found in biological systems was impeded by the lack of knowledge of the functional anatomy and biochemistry of the olfactory system and the odor information processing pathway. This scenario changed during the last few decades. Several experimental findings have improved the understanding of the olfactory structure (Haberly 2001; Yokoi et al. 1995; Ressler et al. 1994; Shepherd 1987, 1994; Vassar et al. 1994; Buck and Axel 1991; Haberly and Bower 1989) and coding mechanisms inspiring new solutions that enhance the capability of artificial olfaction systems.

In this context, a European project, NEUROCHEM, was funded with the aim of developing a novel computing paradigm for chemical sensing based on the information processing performed in the olfactory pathways in the brain, and also of building a large chemical sensor array for generating realistic data inputs for the models. This chapter describes the part of the outcome of the NEUROCHEM project focused on developing and testing the sensor array. The research planned to fabricate a huge number of chemical sensors, 65,536, of tens of different types, exhibiting broad and overlapping sensitivities and different dynamic ranges. Hence, the array has to implement key features of the biological system, as the receptor redundancy, which are expected to contribute to the achievement of an artificial olfactory device having unique performance and able to generate useful data at a speed comparable to biological olfaction to feed into computational models.

3.2 ARTIFICIAL OLFACTORY SYSTEMS

The first electronic nose (e-nose) was built in 1982 by Persaud and Dodd (1982). It was composed of three gas sensors, and its architecture is implemented practically the same in all following systems. From that original device, the expression "electronic nose" was used in conferences and papers, and the initial definition of the term was refined by Gardner and Bartlett (1994): "An electronic nose is an instrument which comprises an array of electronic chemical sensors with partial specificity and an appropriate pattern recognition system, capable of recognising simple or complex odours." By means of this structural design, when a chemical environment is analyzed by an electronic nose, its chemical pattern is transduced in a pattern of electronic signals, the responses of the sensors. Then, this first encoding of the odor is transmitted to a computer, recorded, and processed with a data analysis technique to obtain odor recognition and classification (see Figure 3.1). The analysis of e-nose data is necessary in order to correlate the sensor response to the chemical pattern, as the brain does.

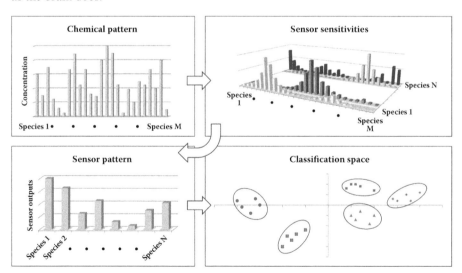

FIGURE 3.1 The principles of an "electronic nose" are shown. The chemical pattern of an environment consisting of a complex mixture of chemicals (top left) is transduced by an array of sensors having broad, overlapping selectivities and sensitivities to a range of chemicals (top right) to produce a sensor response pattern (bottom left). This is processed to allow discrimination and classification of different environmental patterns (bottom right).

The chemical sensors useful for the electronic noses must meet a number of criteria: They must display broad selectivity and high sensitivity to different analytes, as well as rapid response and recovery times. They should be robust, stable, and able to be fabricated reproducibly. Since electronic noses are often proposed for use in poorly controlled environments, sensors with low sensitivity to environmental variables, such as temperature and humidity, are an advantage for these devices. This is not a compulsory feature, because a good delivery system should be able to control all the physical variables that may interfere with the sensor behavior, guaranteeing that their response is only related to the odor. Other desirable requirements for the sensing elements are small size and low power consumption.

A typical gas sensor is composed of two main units: the sensing material, or chemical interactive material (CIM), which "catches" the odorant molecules and reacts, changing one or more of its physical properties, and the transducer, which senses these changes and translates them into an electronic signal. The interactive material is the sensor's heart. It is the interface between the environment and the transducer. The choice of the type of the interactive material to use in an e-nose is based primarily on its sensitivity and selectivity (Di Natale et al. 1998).

Currently, many types of sensing materials can be used for the realization of an electronic nose and a possible list of them is reported below:

- Conductive and nonconductive polymers
- Molecular films
- Catalytic metals
- Metal oxides
- Solid-state ionic conductor

The transducer part of the sensor translates the physical variations that happen in the CIM into an electronic signal, which is suitable for further processing. As there are different transduction mechanisms, the same sensing property of a CIM can be transduced into different electronic signals. Hence, diverse sensors can be built by varying either the CIM or the transducer. Table 3.1 is a list of representative sensors based on different transduction mechanisms.

Although no single sensor technology yet fulfills all of the criteria described above for having a fully featured artificial olfactory system, several electronic noses have been developed during the last 30 years, based on different technological implementations, as depicted in Figure 3.2. Most of them use one sensor technology only, but in order to achieve as much as possible the requirements needed for an electronic nose, hybrid instruments employing several sensor types are appearing currently. Since the introduction of the artificial olfaction, great progress has been made using this approach, and such systems have been applied successfully in many areas, for example, in the industry, for controlling the production chain, and in environmental and food freshness control. Their potential in the medical field as an early and non-invasive diagnostic tool has also been demonstrated (D'Amico et al. 2008a, 2008b; Dutta and Dutta 2006; Dutta et al. 2002, 2005, 2006; Di Natale et al. 2003; Boilot et al. 2002; Gardner et al. 1998).

TABLE 3.1

Types of Sensors Commonly Used

Sensor Type	Operational Principle	References
ChemFET, light addressable potentiometric sensors	Work function	Bratov et al. 2010; Zhang et al. 2000; Morita et al. 1996; Lundstrom et al. 1975
Chemoresistors	Conductivity	Ivanov et al. 2004; Capone et al. 2000; Nanto et al. 2000; Gardner and Bartlett 1991; Kirner et al. 1990; Kurosawa et al. 1990; Bartlett and Ling-Chung 1989
Amperometric gas sensors	Ionic current	Winquist et al. 2000; Chang et al. 1993; Cao et al. 1992
Chemocapacitors	Permittivity	Koll et al. 1998
Thermopile, pellistor catalytic sensor	Temperature	Zemel 1996
Colorimeter, spectrophotometer	Optical spectrum	Ballantine et al. 1992
Optical fibers (fluorescence)	Fluorescence	White et al. 1996
Optical fibers, surface plasmon resonance	Refraction index	Nelson et al. 1996; Chadwick and Gal 1993
Cantilevers, surface acoustic wave, quartz crystal microbalance	Mass	Santonico et al. 2008; Hamacher et al. 2003; Di Natale et al. 1997

Despite the success of this technology in the odor analysis, the artificial systems that have been so far used are distant from mimicking the performance of biological olfaction. Their main limitation is the limited number of sensing materials and sensing elements. Current chemical sensor arrays, either homogeneous or heterogeneous, have few sensing elements when compared to natural noses, commonly from about 10 sensors to about 32 sensors. Many attempts to increase the sensor number and their diversity have been made during the last years, but there are many limiting factors that have not yet permitted us to achieve this objective. In fact, a large array leads to an unmanageable number of electrical connections, high power consumption, long response times, which is something inadmissible in the artificial olfaction, large instruments, and as a consequence of all the previous factors, the reduction of the portability of the system. A recent work appearing in 2009 (Taylor et al. 2009) reported the fabrication of a device containing three sensor arrays of 300 elements using 24 different sensitive materials, still far from the dimensionality and the variety exhibited by the ORNs. We can say that there is a parallelism between human and artificial olfaction, but the performance of the biological system is superior to that of any existing artificial device, as it has not been so far possible to construct apparatus with the same receptors' features and able to perform the same information processing carried out in the brain.

In order to explore the advantages provided to the biological system by the high level of redundancy and the massive convergence of the ORNs to the OB, the

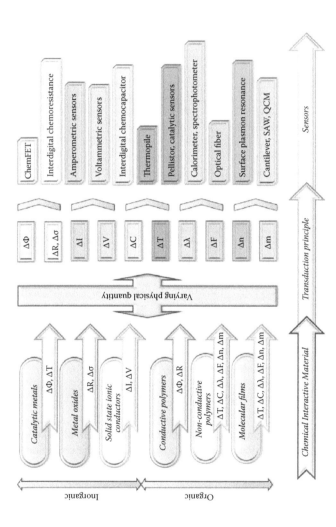

FIGURE 3.2 Technology diagram of existing electronic nose systems. A sensor is composed of two main parts, the chemical interactive material (CIM), which is the element sensitive to the odorant molecules, and the transducer. The CIM changes one or a few physical features as a reaction to the interaction with the odor. The transducer reads the physical variation of the CIM and converts it into an electronic signal. Different CIMs and different transducers can be used, giving rise to different sensors. The CIM commonly employed in the e-noses and their physical characteristics that can be taken as sensor responses are reported in the first column of blocks. The transduction principles normally implemented in the sensors are in the second column. The different sensor technologies obtained by combining CIMs and transducers are in the third column. So, for example, the conductive polymers can be utilized as sensitive material in ChemFETs, where the change of their conductivity, as a consequence of the interaction with an analyte, is detected as a current passing through a transistor, or in chemoresistance sensors, where their electrical resistance is detected.

NEUROCHEM project planned to make a large chemical sensor array consisting of up 2^{16} (65,536) elements, which had to be interfaced with the neuromorphic models developed. Such an array makes it possible to test, for the first time, these types of computational models with realistic inputs. In fact, as the biological olfactory system has a very large number of sensory inputs, the large array creates a wide-ranging set of data to feed into the models. Before our achievements, there were no sensor systems able to achieve such a challenging target.

3.3   CONCEPT AND SYSTEM ARCHITECTURE

One of the objectives of NEUROCHEM was to develop, realize, test, and gather experimental data from an artifact that should carry out the same function as the sensory epithelial layer in the biological system. Hence, our main goal was the implementation of similar features exhibited by the epithelium—a large number of ORNs of multiple types—with broad and overlapping specificity and different dynamic ranges of responses to the same analytes. The sensor array technology chosen consisted of organic conducting polymers that, ideally, allow an infinite number of sensor configurations to be made with broad but overlapping selectivity to different families of chemicals. We have identified and synthesized nearly 30 conductive polymers with broad and overlapped specificity to different volatile organic compounds that are repeatedly deposited over each of the 16 array modules for a total of up to $N = 2^{16}$ (64k) sensing elements. Hence, a high degree of redundancy is implemented, which approaches the redundancy found in biological olfactory systems. For example, in the human olfactory system there are between 10^7 and 10^8 receptors of 350 different types (Buck 2004; Lancet 1991), while fewer are found in simpler animals.

Resistive sensors can be read with several techniques, but the voltage divider is the simplest and allows sequential readout of a multiplicity of elements through a low-impedance analog multiplexer. However, when the sensing elements exceed a few hundreds, they cannot be read by the usual $2N$ leads (or $N + 1$ in the case of a common electrode), as this number of external connections becomes a prohibitive pad-limited problem. Hence, we have conveniently arranged the elements in a matrix configuration having n rows and m columns, with $N = n{\bullet}m$. In this way each resistive element in the array shares its first terminal with all the elements belonging to the same row and the second terminal with the elements in the same column. The total number of wiring connections in this configuration is only $n + m$, with a minimum of $2\sqrt{N}$ for a square array. At a first thought, one may think of reading the resistance value R_{ij} of the generic element belonging to the ith row and jth column using analog multiplexing techniques. One multiplexer connects the jth column to the power source voltage, V_{CC}, through a load resistor, R_{LOAD}, and the other one connects the ith row to ground. Hence, the resistance R_{ij} could be read through the voltage divider between R_{LOAD} and R_{ij}. However, parasitic paths are established in parallel to R_{ij} through multiple parallel combinations of the series of resistors on semiselected row ($l \neq i, j$), semiselected column ($i, k \neq j$), and nonselected rows and columns ($l \neq i, k \neq j$) (see Figure 3.3). These parasitic paths make the output voltage depending not only on R_{ij}, but also on semiselected and nonselected sensor resistances.

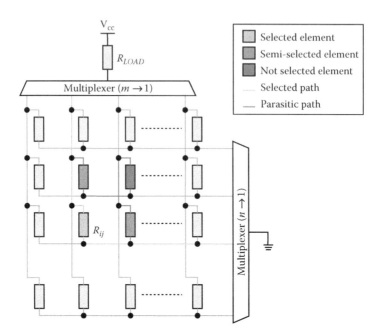

FIGURE 3.3 **(See color insert.)** The matrix configuration adopted for the sensor array reduces the number of electrical wires needed for the sensor connections. The readout circuit is more complex and powers the sensors, while compensating for the electrical cross talk. For element R_{ij}, the column multiplexer (top), selects the jth column to provide power through R_{LOAD}. Simultaneously, the row multiplexer selects the ith row and connects it to ground, so that the green path is selected. All the elements of the jth column are powered, while all the elements of the ith row are connected to the ground. There are semiselected tracks (orange) and nonselected paths (red) that result in unwanted resistances connected in parallel to R_{ij}.

3.4 CROSS TALK COMPENSATION

To eliminate cross talk occurring through unwanted parallel electrical paths, the nonselected rows are forced to the output voltage by the analog feedback buffers (Beccherelli et al. 2009, 2010; Beccherelli and Di Giacomo 2006; Purbrick 1981). In this way, all the sensors belonging to the selected column and to the nonselected rows have a zero-voltage difference across them, and thus whatever their resistance is, no current flows into the unwanted paths. These semiselected column elements appear collectively as a virtual open circuit in parallel to the element being read. Hence, no current flows in any of the parasitic paths exemplified in Figure 3.4, and the scanning circuit behaves as an ideal voltage divider, where one sensor at a time is sequentially connected to the output node and ground.

The multiplexing algorithm works quite well with very large arrays, provided the overall leakage current flowing in off-state ports of the multiplexer remains well below the current flowing in the measurand resistor. However, when acquisition time

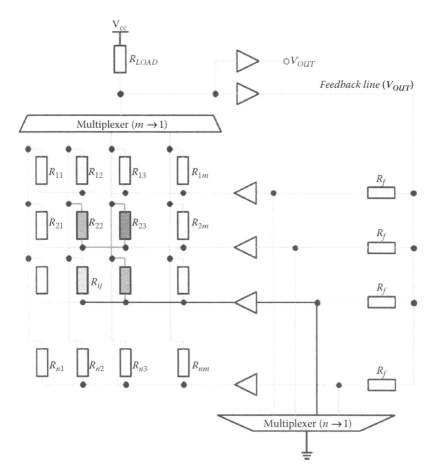

FIGURE 3.4 **(See color insert.)** Simplified scheme of the electronic circuit developed for scanning and reading the large sensor array. In order to read the element R_{ij}, the column multiplexer selects the jth column and connects it to the power through the load resistance. The row multiplexer connects the ith row to ground, while all the other nonselected rows are connected to the output voltage by means of feedback buffers. Hence, the selected sensor is correctly powered, while no current can pass through the nonselected resistances, because of the zero-voltage difference across them. This is illustrated in the figure, where the semiselected element R_{22} is at V_{out} in its first terminal through the column multiplexer and virtually at the same potential in the second terminal because of the feedback line. In this way no current can flow through that element, disconnecting the entire parasitic path from the element being read.

is important, this is limited by the circuit transient time constant, τ, which is directly proportional to the equivalent parasitic capacitance, C_p, and to $R_L//R_{ij}$, $\tau = C_p$ ($R_L//R_{ij}$). The parasitic capacitance is mainly due to the parasitics associated with the low impedance analog multiplexers and scale linearly with the number of multiplexed lines. Hence, smaller matrices can be read faster. Hence, instead of organizing the

64k array as a single 256×256 array, we organized the 64k array into 16 modules, each comprising 4096 elements arranged in a 64×64 matrix. This provides a four-fold decrease of τ. By reading the 16 modules in parallel, a further 16-fold decrease is obtained, resulting in a 64-fold decrease in readout time. This allows a ≈ 1 frame per second acquisition, a realistic target with real polymers and practical electrode geometries. This modular arrangement also adds some flexibility and resilience to the sensors and eases the replacement of damaged parts.

3.5 FABRICATION OF THE RESISTIVE SENSOR ARRAY DIES

Sensor dies were fabricated in batch on a 4 in. borosilicate wafer (0.5 mm thickness; Bullen, Inc.) by using three photolithographic masks that were designed and fabricated. The process performed is delineated and composed of the following four steps:

1. A first layer of 150 nm of Cr-Au was evaporated on the wafer; the row metallization was patterned by subtractive liftoff technique (mask 1).
2. A layer of about 400 nm of SU8 was spun and patterned over the first metallization (mask 2).
3. A third layer of 500 nm of Cr-Au was evaporated; the column metallization was patterned by subtractive liftoff technique (mask 3).
4. A final layer of about 10 μm of SU8 was spun and patterned all over the sensor array to create a confinement tub for each transducer region (mask 2).

The final covering film improves the placing of the sensing material on the array and intrinsically contributes to decrease the cross talk between adjacent elements by breaking the otherwise continuous polymer film.

The developed process allowed fabrication of seven sensor arrays on a single borosilicate wafer. Dies are separated by a blade.

In each sensor die, elements are not identical, but organized in different sets with different dimensions (220×220 μm, 220×420 μm, 420×220 μm, and 420×420 μm), with different spacings and different levels of interdigitation (spanning from 15 to 540) to better accommodate polymers with different resistivities. The single dies cut from the wafer were bonded to the sensor board by using anisotropic tape. Use of glass as a transparent substrate greatly simplifies the alignment and bonding process. A photograph of the array connected to the board is shown in Figure 3.5, and a diagram of the assembled modules is shown in Figure 3.6.

3.6 POLYMER DEPOSITION

Polymer sensors have low power consumption and do not require any heaters, as they work at room temperature. They exhibit fast response and recovery times and respond to many vapors, including alcohols, esters, aromatic compounds, and alkanes. Their sensitivities and specificities can be tailored by changing the dopant species, adding

FIGURE 3.5 A photograph of the sensor array connected to the sensor board. The binding of the array to the sensor board is made with an anisotropic tape. Two connectors are mounted on the sensor board for the connection of the array with the scanning board.

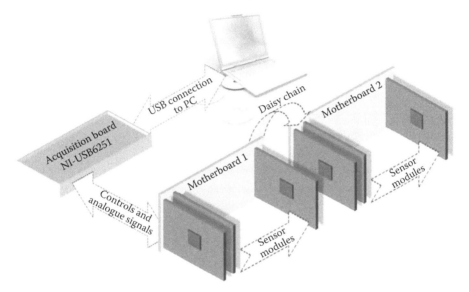

FIGURE 3.6 Overview of the developed scanning and acquisition system, where all the components are sketched. In particular, the system is composed of an acquisition board from National Instruments (model NI 6251-USB), two motherboards, and 16 sensor modules. The acquisition board provides the signals for scanning the sensor array, calculating and storing the optimal load resistance for each sensor, and acquiring the sensor output. Each motherboard provides the power and the ground to eight sensor modules, and routes to the logic units the signal coming from the acquisition board.

functional groups to the main ring, forming copolymers, and incorporating specific reagents, like metals, enzymes, and antibodies, into their structures. Hence, we selected them as sensor coatings for the large-scale sensor array. In particular, it was decided to employ polyanilines, polypyrroles, polythiophenes, and their substitutes, as they showed promise for both their sensing capabilities and ease of tuning.

3.6.1 Conducting Polymers

The interest in conducting polymers as useful materials was initiated in 1977 when MacDiarmid, Heeger, and Shirakawa demonstrated that it was possible to increase the conductivity of polyacetylene by 9 to 13 orders of magnitude, after an oxidation process with iodine (Shirakawa 2001; Shirakawa et al. 1977). These primary experiments, for which the three scientists were awarded the Nobel Prize for Chemistry in 2000 (Heeger 2001; MacDiarmid 2001), produced a great interest in polyacetylene, but the poor chemical stability of this polymer prevented it from achieving great success. Two years later, heterocyclic conducting polymers were discovered (Diaz et al. 1979). They have the same π-system as polyacetylene, giving similar electrical properties. Furthermore, they incorporate a heteroatom into a five-membered ring, resulting in a resonance-stabilized structure when they are polymerized. Hence, they are conductive, just as polyacetylene, but more stable. This event stimulated intense research on these new materials that seemed to offer a combination of properties unlike other known materials (Alan 2001; Heeger and Diaz-Garcia 1998; Salaneck and Brédas 1994; Burn et al. 1992; Brédas et al. 1987; Moraes et al. 1986). Figure 3.7 shows the structure of some commonly used materials that are polymerized to make conducting polymers.

The great advantage of the conducting polymers, also called intrinsically conducting polymers (ICPs) or synthetic metals, is that they combine the good mechanical properties and processability of plastic polymers with electrical, magnetic, and optical properties, similar to those exhibited by the semiconductors. Moreover, they can be easily synthesized by means of electrochemical and chemical processes, they are not expensive, and their morphological, mechanical, and chemical features can be simply tuned. Thanks to these characteristics, they have found great use in electronics as

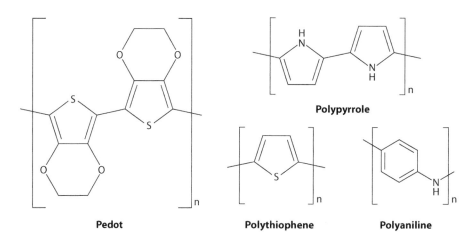

Pedot   Polythiophene   Polyaniline

FIGURE 3.7 Structures of the most common conducting polymers employed in electronics and sensor fields. They are stable heterocyclic compounds offering unique properties. They are able to conduct electricity and have very good mechanical properties as plastic materials.

circuitry components (Cheng et al. 2005), light-emitting diodes (LEDs) (Liu and Cui 2005), organic thin film transistors (TFT) (Kawase et al. 2003), solar cells (Peet et al. 2007), and lasers (McGehee and Heeger 2011). Their capabilities as sensing materials were investigated, for the first time, in 1982 by Nylander et al. (1983). They exposed a sensor based on the heterocyclic polypyrrole to ammonia vapor. Fast and reversible sensor responses, with a linear trend at higher concentrations, were observed. In 1984 the first array of polymer sensors was presented by Persaud and Pelosi (Amrani et al. 1993; Persaud and Pelosi 1985). Currently, organic conducting polymers are largely employed in the chemical sensing field, as both interactive materials and transducers (Basudam and Sarmishtha 2004; Persaud and Pelosi 1985, p. 31).

3.6.2 ELECTRICAL CHARACTERISTICS OF THE CONDUCTING POLYMERS: PRIMARY AND SECONDARY DOPING PROCESS

Conducting polymers have a backbone consisting of alternating single and double bonds, which leads to the formation of π-electrons partially delocalized across a few atoms of the polymer system. The origin of the electrical conductivity of the polymers lies in the presence of these delocalized states that causes, in analogy to inorganic semiconductors, the formation of an energy gap (E_g). The E_g of most conducting polymers is quite large, more than 3 eV wide. Hence, in their undoped state they are insulating. A doping process, performed by redox reactions or, in the case of polyaniline, by protonation, increases the electrical conductivity of the synthetic metals, because it introduces energy levels, polaronic or bipolaronic levels, within the gap.

The initial phase of the doping produces in the system a free radical and a polarization of the structure surrounding the radical. The combination of the charge and the distortion of the backbone is called a polaron. The polaron produces energy states within the gap. With further the doping, the formation of dications by existing polarons is energetically more favorable than the formation of new polarons. Hence, a dication coupled with the distortion of the surrounding lattice is formed and is called a bipolaron. Bipolarons generate electronic states in the band gap that are farther away from the band edges than those generated by polarons. Both, polarons and bipolarons are spread across a few monomer rings, four rings in the case of the polypyrrole.

When an electric field is applied to the system, polarons and bipolarons are able to move along the polymer chain (intrachain charge transport) and between adjacent chains (interchain charge transport). In the intrachain transport the charges move by rearranging the conjugated bonds and the conduction is limited by the conformational and chemical defects present in the chain (Cho et al. 2007). In the intrachain transport, the charge carriers preferably hop or tunnel among chains, and their mobility is limited by the order of the structure.

During the doping procedure, the reduced or oxidized dopant species function as counterions that remain incorporated into the polymer structure to keep the system electrically neutral. The nature of the counterions and their distribution within the chain influence the morphology of the polymer, its electronic, magnetic, and optical properties, and its sensing properties, as it can provide a specific interaction site for analyte gases (Cabala et al. 1997; Pron and Rannou 2002). The doping process just described is defined as primary doping. A secondary dopant can interact with a primary-doped polymer, inducing changes of its physical properties due, principally, to a variation of the molecular conformation (MacDiarmid and Epstein 1994, 1995). In fact, redox sites are spread into the primary doped polymer and define an electron transfer pathway. The electronic coupling between the redox sites is modulated by the different forces among all the atoms in the matrix. The interaction of an external molecule with the polymer modulates the electronic coupling between redox sites and, consequently, the molecular conformation and charge (Janata and Josowicz 2003). Hence, a polymer can be used as a chemical interactive material because an analyte gas acts as a secondary dopant. If the forces involved in the interaction between the polymer and the secondary dopant are not strong, the effect of the dopant is reversible. Moreover, the affinity of a polymer for a secondary dopant can be tuned by using post treatment (Patil et al. 1999), creating specific sites for binding the molecules (Domansky et al. 1997; Josowicz et al. 1999), for example, by means of metallic intrusion and specific primary dopants or by preparing copolymers (Kros et al. 2002).

3.6.3 CONDUCTING POLYMERS IN SENSOR APPLICATIONS

The interaction between conducting polymers and secondary dopants changes, selectively and in a reversible way, the electrical, optical, and magnetic properties of the polymer, giving the possibility to make sensors based on different transducer mechanisms. Polymer sensors exploiting the variation of an electrical resistance, work function, Schottky barrier, or optical spectrum have been developed (Covington et al. 2001; Jin et al. 2001; Karin 2001; Lange and Mirsky 2011; Péres et al. 2012; Rivera et al. 2003). Furthermore, conducting polymers have been used as coating of quartz crystal microbalance and surface acoustic wave sensors. In this case they act as active layers able to interact with the analyte, and none of their electrical or optical properties are detected, but they are utilized to increase the sensitivity and selectivity of the piezoelectric material (Henkel et al. 2001; Hwang et al. 2001; Milella and Penza 1998; Penza et al. 1998).

The interactions between ICPs and molecules in the gas/vapor phase are still not completely understood, because there are multiple types of interactions. It has been demonstrated that redox-active gases such as NH_3, NO_2, I_2, and H_2S interact with polypyrrole, polythiophene, and polyaniline by electron transfers, changing the resistance and work function of the sensitive materials. Electron acceptors, such as NO_2 and I_2, are able to remove electrons from the aromatic rings of the polymers. If this happens to a p-type material, its conductivity is increased; if it happens to an n-type, the conductivity is decreased. Electron donors generate a contrary effect, increasing the conductivity of the n-types and decreasing that of p-type polymers. This effect can be reversed removing or compensating the introduced charge, *dedoping process*. Many different kind of polymeric sensors are based on this interaction (Nguyen and Potje-Kamloth 2001; Xie et al. 2002).

Analytes, showing weaker Lewis acid and base features, give rise to partial electron transfer, which still changes the number of the charge carriers in the polymer backbone and its conductivity. The electronegativity of the vapors and the work function of the polymer determine the direction of such transfer (Blackwood and Josowicz 1991).

Electron transfer can happen on a component mixed with the polymer, on the counterions introduced during the doping process, and on the side chains added to the main ring of the polymer (Chyla et al. 2001; Rungnapa 2003). This makes available more opportunities for tuning the selectivity of the conducting polymer sensors. The introduction of catalysts in the polymer can be used to make sensors able to detect inert gases (Athawale et al. 2006).

Swelling of the polymers, in consequence of the interaction with an analyte, is also employed in chemical sensors. The introduction of gas molecules in the polymer structure changes the morphology of the polymer, and therefore its conductivity, as the interchain distance increases. The swelling is regulated by the vapor molecular volume, the affinity of the vapor to the sensing polymer, and the physical state of the polymer. This mechanism is observable both in a pure conductive polymer and in a composite of it with other filling materials. In the latter case, the two components can swell, making more complex the interaction mechanism (Hosseini and Entezami 2003; Sakurai et al. 2002; Virji et al. 2004).

It is clear that the possible interactions between analytes and ICPs are numerous and not very well understood. Consequently, the interpretation of the sensor signal modulation on exposure to an analyte is difficult. Despite this, conducting polymer sensors are widely used for chemical sensing.

3.6.4 Sensing Material Deposition

As the characteristics of the synthetic metals described above meet our aim of emulating the olfactory epithelium of the biological olfactory system, they were chosen as sensing materials for the large array.

A total of 31 different polymers were employed for our system. They were all polyanilines, polypyrroles, and polythiophenes. These incorporated a variety of dopant species, while adding diverse side chains to the main polymer ring and preparing copolymers ensured diversity in the sensitivity and selectivity to different analytes.

The deposition of the chemically interactive materials onto the sensor substrates was not a simple task. Great limitations were imposed by the matrix arrangement of the sensors and the employment of the conductive tape for bonding the sensors to the scanning board. The anisotropic tape used for bonding necessitated the application of high temperatures. As high temperatures can damage the polymers, they were deposited onto the array after its connection to the electronic circuit. This meant that both the array and the scanning board were involved in the deposition process, hindering the use of many other techniques. In particular, all those methods that require the immersion of the substrates into the polymer solutions, the spreading of the material onto the whole array, or the exposure of the substrate to aggressive chemicals were not practicable. Electropolymerization, one of the most applied polymer deposition methods, was not also possible due to the matrix configuration adopted for the sensors.

3.6.4.1 Airbrush Deposition Process: System Setup and Ink Preparation

The spray casting process was set up into a fume cupboard and performed under air. Pressurized nitrogen was used as carrier gas. A dual-action Paasche airbrush was employed, and it was fixed to a mechanic arm. The sensor array, already bonded to the sensor board, was also fixed to an arm. The positions of both, the array and the gun, could be easily changed. The pressure of the atomizer gas was always kept below 2 bar in order to avoid the possibility of droplets previously placed onto the substrates being blown off by the successive spray steps. Chloroform was the solvent selected for the inks. This choice was because many polymers are partially or completely soluble in it, and there is no need to heat the substrate to aid evaporation of the solvent.

The distance between the substrate and the airbrush was adjusted for each ink in a range from 10 to 20 cm. If the distance between the gun and the target is too high, the solvent evaporates during flight, and dry powder is deposited on the sensors, so the resulting film is inhomogeneous and its resistivity is high. On the contrary, if the distance is too low, when the polymer hits the substrate, it is too wet and the solution flows down. Even in this case, the film is not homogeneous and the thickness is not controlled. At the optimum distance, the polymer is still wet when it reaches the sensors, but the small amount of the residual solvent is just enough to allow merging of the droplets to form a uniform film.

Seven masks were realized, one for each region the array is subdivided into, making it possible to deposit up to seven polymers on each array. The placing of the polymers on a surface comprising more sensors, instead of just one sensor, was not a problem: the thick final layer of SU-8 grown on the sensor array reduced the risk of cross-links between neighboring sensors. The spray of a single polymer was performed in several steps, during which the distance between the target and the gun, the pressure of the atomizer gas and the carrier gas, and the duration of the deposition step are kept constant. Figure 3.8 illustrates a sensor substrate that has been partially deposited with three different materials.

3.7 LABORATORY EXPERIMENTS

Once the array was built, adopting the strategies explained above, laboratory experiments were performed to characterize and test the sensors and validate the neural models developed by the NEUROCHEM partners. A semiautomatic chemical analyte delivery system based on permeation tubes and bubblers coupled to mass flow controllers (MFCs) for vapor dilution was constructed for this aim. It allowed the delivery of pure analytes and binary mixtures over a wide range of concentrations, 1 ppm to several thousands of ppm. Humidity and temperature sensors enabled the monitoring of these physical parameters that may affect the sensor response.

Initial tests were carried out on 4096 sensors and 5 different analytes: 2-butanone, ethanol, ammonia, butyric acid, and acetic acid. They were employed to carry out an initial characterization of the sensor and evaluate the distribution of the sensor resistances implemented in the array. Subsequently, measurements on pure ethanol and 2-butanone were performed on 16,384 sensors and 24 different polymers. The data collected were employed to evaluate the classification and quantification capability of

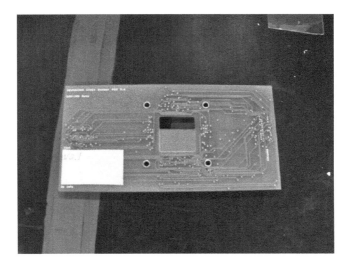

FIGURE 3.8 Photograph of a sensor die with three regions covered with different conducting polymers. By using mechanical masks, it was possible to place up to seven polymers on the sensor array by spray coating. The spray was carried out by using a dual-action Paasche airbrush. The physical and morphological features of the sprayed layers are influenced by parameters such as the distance between the sensor and the gun, the concentration of the inks, the solvent used, and the pressure of the atomizer gas.

TABLE 3.2
Measurement Protocol for Ethanol

Ethanol (Equilibrium Vapor Pressure)

Concentrations (ppm)	Number of Repetitions	Duration of the Measurement Phase (s)
343	10	250–300
686	12	300
1716	8	300–400
3433	8	350
6867	8	400
13,734	8	550

the array and determine concentration ranges, exposure time, and cleaning time of the sensors for the two analytes. These measurements are described in Tables 3.2 and 3.3.

A third series of measurements was carried out on binary mixtures of ethanol and 2-butanone and was used to generate two data sets, data set I and data set II, aimed at determining how the sensors deal with binary mixtures. Data set I captured the sensor responses to two odors and their binary mixtures for three different ratios. Figure 3.9 shows the composition of the five odors (two pure + three mixtures) generated. Starting from pure analyte, A, for instance, the percentage of A in the mixture decreases, while the percentage of odor B in the mixture increases until pure analyte B is reached. With

TABLE 3.3

Measurement Protocol for 2-Butanone

2-Butanone (Equilibrium Vapor Pressure)

Concentrations (ppm)	Number of Repetitions	Duration of the Measurement Phase (s)
488	10	200
976	10	200
1464	10	300
2440	10	300
4880	10	500
9760	10	600
19,520	8	550

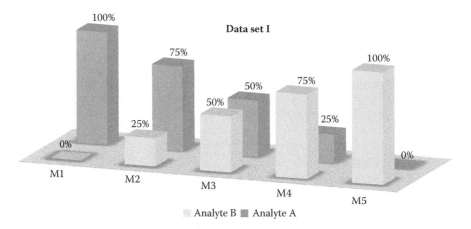

FIGURE 3.9 Histogram representation of data set I. This data set was aimed at capturing the sensor responses to two odors and their binary mixtures for three different ratios. Starting from a pure analyte, A, for instance, the percentage of A in the mixture decreases, while the percentage of the odor B in the mixture increases until pure analyte B is reached. With this data set a collection of odors with increasing similarity to an odor (e.g., B) and increasing dissimilarity to another odor (e.g., A) was obtained.

this data set a collection of odors with increasing similarity to an odor (e.g., B) and increasing dissimilarity to another odor (e.g., A) was obtained. At least eight repetitions were carried out for each concentration step. These measurements were used to assess whether the array is able to segment the binary mixtures into their components.

Data set II was collected to evaluate the capability of the system to recognize the variations in the concentration of a compound in an environment, when a high background of a second compound is present. Hence, it was built by exposing the sensors to different concentrations of one odor, 2-butanone, while a constant background of another compound, ethanol, was present, as described in the histogram in Figure 3.10.

The results of these studies when the data were processed using neuromorphic models of the olfactory system are presented and discussed in Chapter 6 of this

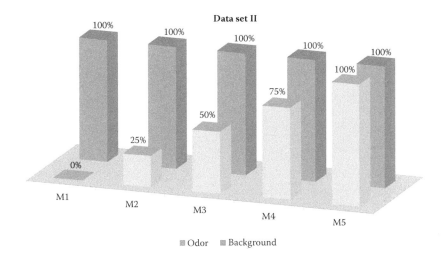

FIGURE 3.10 Histogram representation of data set II. This involved exposing the gas sensors to different concentrations of one odor while a constant high background of another compound was present.

book. Here, the initial analysis of the data using standard techniques of chemometric analysis is reported. This preliminary analysis was aimed at testing the electronics developed for reading the sensors and the capability of the array only, that is, without taking into account any improvement introduced by the neuromorphic computational models.

Before performing the analysis, the sensor signals were preprocessed. An initial low-pass filtering of the signal was performed by means of a moving-average filter with a 30 s window. This process produced a smoother sensor signal, reducing its fluctuations.

Then, the sensor response to an analyte was calculated as the fractional variation of its resistance:

$$\frac{R - R_0}{R_0}$$

where R_0 is the sensor resistance at the beginning of the measurement and R is the resistance at steady state.

The data after preprocessing were subjected to an initial exploratory principal component analysis (PCA) and, in the cases of regression problems, to a partial least-squares (PLS) analysis. The results of such analyses are reported in the following sections.

3.7.1 CHARACTERIZATION OF A PRELIMINARY 4096-SENSOR ARRAY

The measurements on the first 4096 sensors allowed characterization of the sensors in terms of the range of their resistance values. A distribution of the sensor resistances read from the array under a stable baseline was obtained, and it gave an indication about the different specificities and sensitivities of the sensors. The sensors

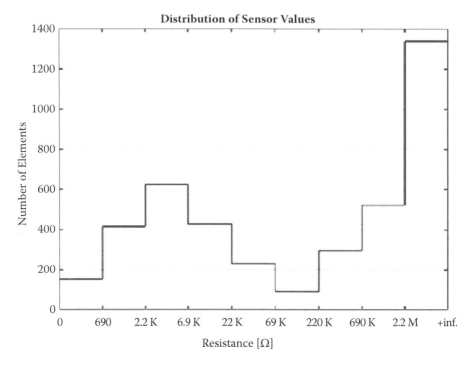

FIGURE 3.11 Ranges of sensor resistance values found in the 4096-element array. The figure depicts the distribution of the values over the nine classes determined by the load resistors.

were classified into nine different groups, resulting from the ranges of adaptation determined by the seven values of the load resistors implemented in the device. Two classes were added—for values below 690 Ω and over 2.2 MΩ—as if two more load resistors were present in the device, with values of 470 Ω and 4.7 MΩ. The sharing of the sensor resistance values over these classes, together with their spatial distribution on the sensor board, is depicted in Figure 3.11. The spread of the sensor resistances—from low to high values—depended on the different materials, the electrode geometries, and the variability associated with the deposition process.

Sensors belonging to different classes showed different sensing behavior, as can be seen in Figure 3.12, where the responses to 3 ppm ammonia are plotted. A rapid visual inspection of these responses showed generally correct behavior, with an increasing rate of noisy sensors in the classes where there were high resistance values. This was due to the presence of faulty sensors (in particular because of defective bonding of some rows and columns of the sensor die) that were read as high resistors by the reading electronics. The observation of the distribution on the sensor array of the elements of the last class confirms this supposition, as most of them are not spread on the array, but are located on whole rows and columns.

From this preliminary study, it was evident that effectively a very wide range of sensor resistances was implemented in the system, and that different groups of sensors exhibited different responses to the same analyte.

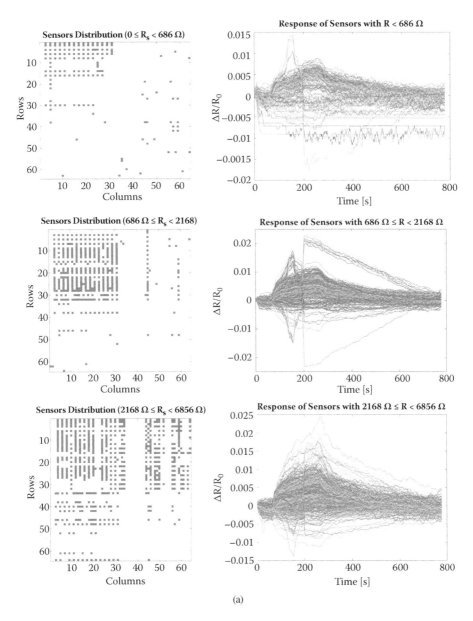

(a)

FIGURE 3.12 Distribution of the sensors, pointing out the classes of resistance values. Close to the spatial distribution is depicted the response of the corresponding sensors to a 3 ppm pulse of ammonia. While there is generally correct behavior of the sensor array, it is possible to notice a certain number of unresponsive or noisy sensors, especially those belonging at higher resistance values. This is to be attributed in particular to faulty sensors, where the bonding failed to correctly connect rows or column connections. In the last series of images (b), corresponding to sensors read as having a resistance value greater than 2.2 MΩ, a clear pattern of defects-matching rows and columns is evident.

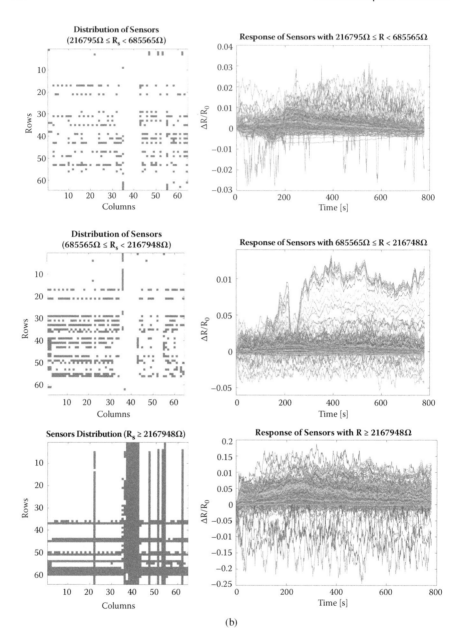

(b)

FIGURE 3.12 (continued) Distribution of the sensors, pointing out the classes of resistance values. Close to the spatial distribution is depicted the response of the corresponding sensors to a 3 ppm pulse of ammonia. While there is generally correct behavior of the sensor array, it is possible to notice a certain number of unresponsive or noisy sensors, especially those belonging at higher resistance values. This is to be attributed in particular to faulty sensors, where the bonding failed to correctly connect rows or column connections. In the last series of images (b), corresponding to sensors read as having a resistance value greater than 2.2 MΩ, a clear pattern of defects-matching rows and columns is evident.

3.7.2 CLASSIFICATION AND QUANTIFICATION

From the initial characterization of 4096 sensors, four sensor modules were combined to produce 16,384 sensors. The ability of the final system composed of 16,384 elements to classify and quantify pure analytes was evaluated by processing the data collected with the measurement protocol detailed in Tables 3.1 and 3.2. First, a principal component analysis was carried out, generating the score plots shown in Figures 3.13 and 3.14 for ethanol and 2-butanone, respectively.

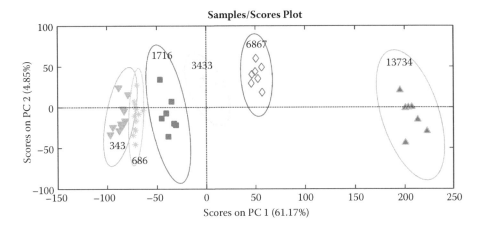

FIGURE 3.13   The score plot shown in this figure was achieved by applying the principal component analysis to the data generated by exposing the large array of 16,384 sensors to ethanol vapor at the different concentrations listed in Table 3.2. Information about the concentration is clearly contained in the first principal component.

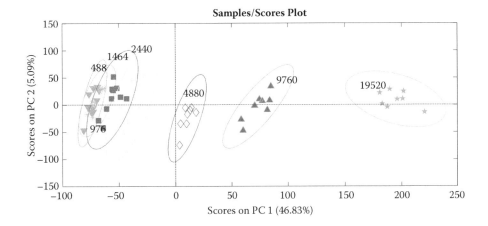

FIGURE 3.14   Principal component analysis applied to the data generated by exposing the large array of 16,384 sensors to 2-butanone vapor at the different concentrations listed in Table 3.3.

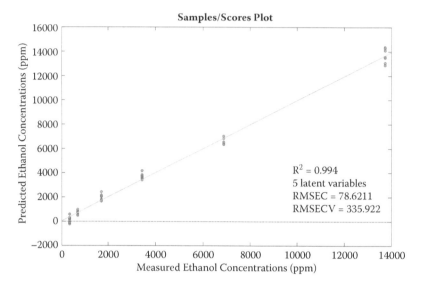

FIGURE 3.15   This figure shows the score plot achieved with a partial least-squares regression model built for predicting the ethanol concentrations. The model was created by using five latent variables, which is the optimum number that minimizes the root mean square error of the cross-validation method used (RMSECV). The root mean square error of the calibration (RMSEC) and RMSECV obtained by the model were 336 and 79 ppm, respectively. These values are smaller than those calculated for 2-butanone, confirming that the large-scale array can discriminate the ethanol better than the 2-butanone, as suggested by the unsupervised analysis.

In both cases, the first principal component alone carries the information about the analyte concentration. Hence, from the simple observation of the first principal component it is possible to discriminate between the different odor intensities. Nevertheless, the clusters pointed out by the score plots are clearly separated at higher concentrations, while there is a degree of overlap at the lower concentrations, in particular for 2-butanone. For ethanol, a small overlap exists between the clusters related to the first two concentrations, while the remaining concentrations are noticeably separated.

As a result of this exploratory analysis, it was possible to build two partial least-squares regression models to predict the different concentrations of ethanol and 2-butanone. Their prediction abilities were evaluated by means of the leave-one-out cross-validation technique. The two models are shown in Figure 3.15 for ethanol and Figure 3.16 for 2-butanone.

Both plots exhibit a good linear relation between the delivered analyte concentrations and the predicted concentrations. If we look at the errors produced by the models on the calibration sets (RMSEC) and the validation sets (RMSECV), those regarding ethanol, which are 79 and 336 ppm, respectively, are quite lower than those of 2-butanone, which are 779 and 845 ppm, respectively. Hence, we can conclude the system is able to quantify both analytes, but with different performances

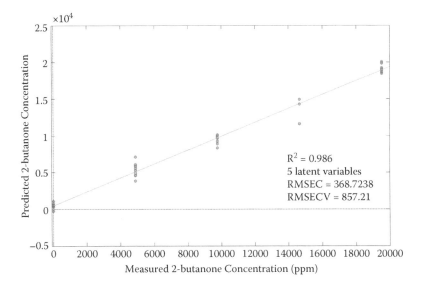

FIGURE 3.16 A partial least-squares regression model, able to predict the different concentrations of 2-butanone, was built by considering the first two latent variables. This is the number of LVs that minimizes the RMSECV. The errors from the calibration set and validation set using this model are 779 and 845 ppm, respectively, validating the results achieved with the PCA.

for each. In particular, ethanol is detected with a limit of detection lower than 2-butanone and with better resolution.

3.7.3 BINARY MIXTURES, SEGMENTATION, AND BACKGROUND SUPPRESSION

After verifying that the sensor system is able to detect and predict different concentrations of pure analytes, we checked its behavior when exposed to binary mixtures. Also in this case, an initial exploratory principal component analysis was applied to data set I together with the data of pure ethanol and 2-butanone at different concentrations. The resulting score plot is shown in Figure 3.17. It shows that the mixtures are correctly recognized as different from the pure compounds.

Moreover, with the increase in concentration of one analyte and the decrease in the concentration of the other component in the mixtures, the similarity of the mixture to the first compound is enhanced, while that to the second compound diminishes. This is correctly represented in the score plot by the distance between the clusters. The clusters formed by mixtures with lower concentrations of 2-butanone and higher concentrations of ethanol are closer to the clusters composed of pure ethanol and farther away from those composed of pure 2-butanone. On the contrary, the mixtures with lower concentrations of ethanol and higher concentrations of 2-butanone give clusters that are closer to the groups of measurements relative to pure 2-butanone and farther away from those of pure ethanol. Looking at the

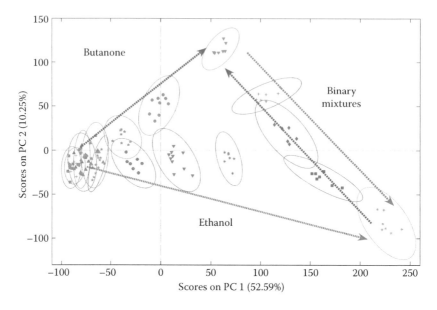

FIGURE 3.17 The score plot presented in this figure was achieved by performing the principal component analysis on the data set I and on the data regarding the pure analytes, in order to check how the large-scale array deals with binary mixtures. Pure 2-butanone projects from the left upwards have increasing concentrations. Pure ethanol projects from the left to right are along the bottom. The binary mixture projects from pure 2-butanone are on the top left of the figure, down to pure ethanol on the bottom right (as shown by the arrows), with the concentration ratios shown in Figure 3.9. The mixtures are correctly recognized as different from the pure compounds. Furthermore, the clusters more similar to ethanol are close to those formed by pure ethanol and far from those containing pure 2-butanone, while the cluster formed by mixtures more similar to 2-butanone are close to those of pure 2-butanone and far from those containing pure ethanol. This score plot reveals that the system should be able to discriminate one analyte in a binary mixture.

different concentrations of the components in the mixtures, clustering determined by the concentrations is identifiable in the score plot, indicating that the large array should be able to recognize not only the presence of a single analyte in a binary mixture, but also its concentration. Hence, odor segmentation processes seem to be possible with the developed array.

To test the capability of the array in terms of background suppression, data set II was analyzed together with the sensor responses to pure butanone (without ethanol background). The score plot achieved by employing principal component analysis is shown in Figure 3.18. In the plot, the clusters generated by different concentrations of pure 2-butanone are different from those generated changing the concentration of 2-butanone when the background of ethanol is present. This indicates that the system correctly detects the background. The first principal component carries the information of the presence of the ethanol. However, despite the strong background, the sensors are able to discriminate dissimilar concentrations of 2-butanone, and this information is given by the second principal component.

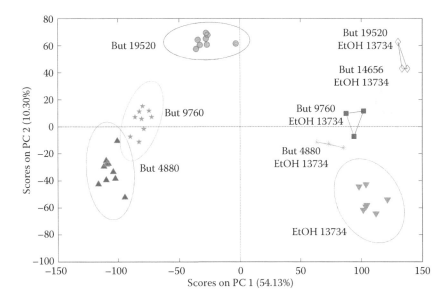

FIGURE 3.18   The score plot shown in this figure was built by performing principal component analysis on the measurements of 2-butanone with and without the background. The first principal component discriminates the presence of the background, while the second principal component maintains the information on the different concentrations of 2-butanone.

When a PLS regression was carried out, the plot in Figure 3.19 was generated. The resulting regression curve shows excellent linearity and an error in cross-validation similar to that obtained with pure 2-butanone (without ethanol background), meaning that the predictive model is not degraded by the presence of the background. These results from data set II indicate that it is possible to train the sensing device to discriminate different concentrations of one single analyte even in the presence of a strong background, suppressing it.

3.8   CONCLUSION

We have developed a portable large-scale chemical sensor array that emulates some interesting characteristics of biological sensory systems. The modular approach adopted for reading the sensors allows data to be acquired in parallel from up to 16 sensor arrays, each composed of 4096 elements, with an acquisition rate of about 1 frame per second. Organic conducting polymers function as sensing materials showing fast response and recovery times at room temperature.

This preliminary analysis by conventional PCA and PLS was aimed to test the electronics developed for reading the sensors and the capability of the array only, that is, without taking into account any improvement that may be introduced by applying neuromorphic computational models. The results show that the system has different specificities and sensitivities to different compounds, and it is able to deal with binary mixtures. Looking at the different concentrations of the components in

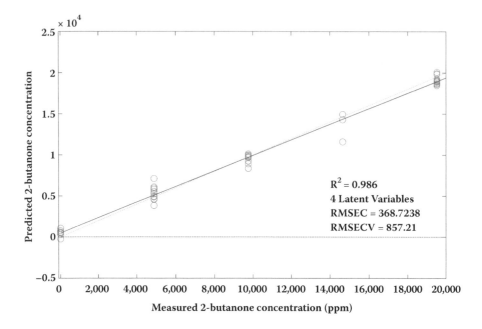

FIGURE 3.19   Partial least-squares analysis on data set II together with the data from pure 2-butanone (2-butanone variations with and without ethanol background). The resulting RMSECV remains at the same level of that achieved by the regression model built for predicting the concentrations of pure 2-butanone. It seems that the predictive model is not degraded by the presence of the background. This result indicates that it is possible to train the sensing device to discriminate different concentrations of one single analyte even in the presence of a high background of another analyte.

the mixtures, results show that the large array can recognize not only the presence of a single analyte in a binary mixture, but also its concentration, thus allowing odor segmentation. Regression analysis shows excellent linearity and an error in cross-validation similar to that obtained with a pure analyte, meaning that the predictive model is not degraded by the presence of the background. Results clearly indicate that it is possible to train the sensing device to discriminate different concentrations of one single analyte even in the presence of a high background.

The developed system is unique with such a large number of sensors and high sensing variability. Hence, it is possible, for the first time, to start testing computational hypotheses appropriate to biological chemosensory systems with realistic input signals for the computational models.

ACKNOWLEDGMENT

This work was supported through funding from the EU project Neurochem (FP7-ICT FET Project 216916).

REFERENCES

Alan, J. 2001. Semiconducting and metallic polymers: The fourth generation of polymeric materials. *Synthetic Metals* 125 (1): 23–42. http://www.sciencedirect.com/science/article/pii/S0379677901005094.

Amrani, M. E. H., M. S. Ibrahim, and K. C. Persaud. 1993. Synthesis, chemical characterisation and multifrequency measurements of polyN-(2-pyridyl) pyrrole for sensing volatile chemicals. *Materials Science and Engineering C* 1: 17–22.

Athawale, A., S. V. Bhagwat, and P. P. Katre. 2006. Nanocomposite of Pd-polyaniline as a selective methanol sensor. *Sensors and Actuators B: Chemical* 114 (1): 263–267.

Ballantine Jr., D. C., G. J. Maclay, and J. R. Stetter. 1992. An optical waveguide acid vapor sensor. *Talanta* 39 (12): 1657–1667. http://www.sciencedirect.com/science/article/pii/003991409280200W.

Bartlett, P. N., and S. K. Ling-Chung. 1989. Conducting polymer gas sensors. Part III. Results for four different polymers and five different vapours. *Sensors and Actuators* 20 (3): 287–292. http://www.sciencedirect.com/science/article/pii/0250687489801271.

Basudam, A., and M. Sarmishtha. 2004. Polymers in sensors application. *Progress in Polymer Science* 29: 699–766.

Beccherelli, R., and G. Di Giacomo. 2006. Low cross-talk readout electronic system for arrays of resistive sensors. Paper presented at Eurosensors XX, Göteborg, Sweden, September 17–20, 2006.

Beccherelli, R., E. Zampetti, S. Pantalei, M. Bernabei, and K. Persaud. 2009. Very large chemical sensor array for mimicking biological olfaction. In *Proceedings of the 13th International Symposium on Olfaction and Electronic Nose*, 155–158. WOS: 000268929400034.

Beccherelli, R., E. Zampetti, S. Pantalei, M. Bernabei, and K. Persaud. 2010. Design of a very large chemical sensor system for mimicking biological olfaction. *Sensors and Actuators B—Chemical* 146 (2): 446–452. WOS: 000277902400006.

Blackwood, D., and M. Josowicz. 1991. Work function and spectroscopic studies of interactions between conducting polymers and organic vapors. *J. Phys. Chem.* 95 (1): 493–502.

Boilot, P., E. L. Hines, J. W. Gardner, R. Pitt, S. John, J. Mitchell, and D. W. Morgan. 2002. Classification of bacteria responsible for ENT and eye infections using the Cyranose system. *IEEE Sensors Journal* 2 (3): 247–253.

Bratov, A., N. Abramova, and A. Ipatov. 2010. Recent trends in potentiometric sensor arrays—A review. *Analytica Chimica Acta* 678 (2): 149–159. WOS: 000283388000003.

Brédas, J. L., F. Wudl, and A. J. Heeger. 1987. Polarons and bipolarons in doped polythiophene: A theoretical investigation. *Solid State Communications* 63 (7): 577–580. http://www.sciencedirect.com/science/article/pii/0038109887908568.

Buck, L., and R. Axel. 1991. A novel multigene family may encode odorant receptors: A molecular basis for odor recognition. *Cell* 65 (1): 175–187. http://linkinghub.elsevier.com/retrieve/pii/009286749190418X.

Buck, L. B. 2004. Olfactory receptors and odor coding in mammals. *Nutrition Reviews* 62: S184–S188. http://dx.doi.org/10.1111/j.1753–4887.2004.tb00097.x.

Burn, P. L., A. B. Holmes, A. Kraft, D. D. C. Bradley, A. R. Brown, and R. H. Friend. 1992. Synthesis of a segmented conjugated polymer chain giving a blue-shifted electroluminescence and improved efficiency. *Journal of the Chemical Society Chemical Communications* (1): 32–34. http://dx.doi.org/10.1039/C39920000032.

Cabala, R., V. Meister, and K. Potje-Kamloth. 1997. Effect of competitive doping on sensing properties of polypyrrole. *J. Chem. Soc., Faraday Trans.* 93 (1):131–137.

Cao, Z., W. J. Buttner, and J. R. Stetter. 1992. The properties and applications of amperometric gas sensors. *Electroanalysis* 4 (3): 253–266. http://dx.doi.org/10.1002/elan.1140040302.

Capone, S., P. Siciliano, F. Quaranta, R. Rella, M. Epifani, and L. Vasanelli. 2000. Analysis of vapours and foods by means of an electronic nose based on a sol-gel metal oxide sensors array. *Sensors and Actuators B—Chemical* 69 (3): 230–235. http://www.sciencedirect. com/science/article/pii/S0925400500004962.

Chadwick, B., and M. Gal. 1993. Enhanced optical detection of hydrogen using the excitation of surface plasmons in palladium. *Applied Surface Science* 68 (1): 135–138. http://www. sciencedirect.com/science/article/pii/016943329390222W.

Chang, S. C., J. R. Stetter, and C. S. Cha. 1993. Amperometric gas sensors. *Talanta* 40 (4): 461–477. http://www.sciencedirect.com/science/article/pii/0039914093800029.

Cheng, K., M. H. Yang, W. W. W. Chiu, C. Y. Huang, J. Chang, T. F. Ying, and Y. Yang. 2005. Ink-jet printing, self-assembled polyelectrolytes, and electroless plating: Low cost fabrication of circuits on a flexible substrate at room temperature. *Macromolecular Rapid Communications* 26 (4): 247–264. http://dx.doi.org/10.1002/marc.200400462.

Cho, S. H., K. T. Song, and J. Y. Lee. 2007. Recent advances in polypyrrole. In *Handbook of Conducting Polymers,* ed. T. A. Skotheim and J. R. Reynolds, 8.1–8.87. Boca Raton, FL: CRC Press.

Chyla, A., A. Lewandowska, J. Soloducho, A. Gorecka-Drzazga, and M. Szablewski. 2001. 4-t-butyl-CuPc-PODT molecular composite material for an effective gas sensor. *IEEE Transactions on Dielectrics and Electrical Insulation* 8 (3): 559–565.

Covington, J. A., J. W. Gardner, D. Briand, and N. F. de Rooij. 2001. A polymer gate FET sensor array for detecting organic vapours. *Sensors and Actuators B: Chemical* 77 (1–2): 155–162.

D'Amico, A., R. Bono, G. Pennazza, M. Santonico, G. Mantini, M. Bernabei, M. Zarlenga, C. Roscioni, E. Martinelli, R. Paolesse et al. 2008a. Identification of melanoma with a gas sensor array. *Skin Research and Technology* 14 (2): 226–236.

D'Amico, A., C. Di Natale, R. Paolesse, A. Macagnano, E. Martinelli, G. Pennazza, M. Santonico, M. Bernabei, C. Roscioni, G. Galluccio, R. Bono, E. F. Agrò, and S. Rullo. 2008b. Olfactory systems for medical applications. *Sensors and Actuators B—Chemical* 130 (1): 458–465. http://www.sciencedirect.com/science/article/pii/S092540050700723X.

Di Natale, C., A. Macagnano, R. Paolesse, E. Tarizzo, A. D'Amico, F. Davide, T. Boschi, M. Faccio, G. Ferri, F. Sinesio, F. M. Bucarelli, E. Moneta, and G. B. Quaglia. 1997. A comparison between an electronic nose and human olfaction in a selected case study. *Transducers '97 Proceedings* 2: 1335–1338.

Di Natale, C., A. Macagnano, E. Martinelli, R. Paolesse, G. D'Arcangelo, C. Roscioni, A. Finazzi-Agrò, and A. D'Amico. 2003. Lung cancer identification by the analysis of breath by means of an array of non-selective gas sensors. *Biosensors and Bioelectronics* 18 (10): 1209–1218. http://www.sciencedirect.com/science/article/pii/S0956566303000861.

Di Natale, C., A. Macagnano, G. Repole, G. Saggio, A. D'Amico, R. Paolesse, and T. Boschi. 1998. The exploitation of metalloporphyrins as chemically interactive material in chemical sensors. *Materials Science and Engineering* C 5 (3–4): 209–215. http://www. sciencedirect.com/science/article/pii/S0928493197000453.

Diaz, A. F., K. K. Kanazawa, and G. P. Gardini. 1979. Electrochemical polymerization of pyrrole. *Chemical Communications* 14: 635–636.

Domansky, K., J. Li, and J. Janata. 1997. Selective doping of chemically sensitive layers on a multisensing chip. *Journal of the Electrochemical Society* 144 (4): L75–L78.

Dutta, R., A. Das, N. G. Stocks, and D. Morgan. 2006. Stochastic resonance-based electronic nose: A novel way to classify bacteria. *Sensors and Actuators B—Chemical* 115 (1): 17–27. WOS: 000236929100004.

Dutta, R., and R. Dutta. 2006. "Maximum probability rule" based classification of MRSA infections in hospital environment: Using electronic nose. *Sensors and Actuators B— Chemical* 120 (1): 156–165. WOS: 000241489300022.

Dutta, R., E. Hines, J. Gardner, and P. Boilot. 2002. Bacteria classification using Cyranose 320 electronic nose. *Biomedical Engineering Online* 1 (1): 4. http://www.biomedical-engineering-online.com/content/1/1/4.

Dutta, R., D. Morgan, N. Baker, J. W. Gardner, and E. L. Hines. 2005. Identification of *Staphylococcus aureus* infections in hospital environment: Electronic nose based approach. *Sensors and Actuators B—Chemical* 109 (2): 355–362. http://www.sciencedirect.com/science/article/pii/S0925400505000250.

Gardner, J. W., and P. N. Bartlett. 1991. Potential applications of electropolymerized thin organic films in nanotechnology. *Nanotechnology* 2 (1): 19. http://stacks.iop.org/0957-4484/2/i=1/a=003.

Gardner, J. W. and P. N. Bartlett. 1994. A brief history of electronic noses [abstract]. *Sensors and Actuators B—Chemical* 18 (1–3): 210.

Gardner, J. W., M. Craven, C. Dow, and E. L. Hines. 1998. The prediction of bacteria type and culture growth phase by an electronic nose with a multi-layer perceptron network. *Measurement Science and Technology* 9 (1): 120–127. http://dx.doi.org/10.1088/0957-0233/9/1/016.

Haberly, L. B. 2001. Parallel-distributed processing in olfactory cortex: New insights from morphological and physiological analysis of neuronal circuitry. *Chemical Senses* 26 (5): 551–576. http://chemse.oxfordjournals.org/content/26/5/551.abstract.

Haberly, L. B., and J. M. Bower. 1989. Olfactory cortex: Model circuit for study of associative memory? *Trends in Neurosciences* 12 (7): 258–264. http://www.sciencedirect.com/science/article/pii/0166223689900258.

Hamacher, T., J. Niess, P. Schulze Lammers, B. Diekmann, and P. Boeker. 2003. Online measurement of odorous gases close to the odour threshold with a QMB sensor system with an integrated preconcentration unit. *Sensors and Actuators B—Chemical* 95 (1–3): 39–45. http://www.sciencedirect.com/science/article/pii/S0925400503004003.

Heeger, A. J. 2001. Nobel lecture: Semiconducting and metallic polymers: The fourth generation of polymeric materials. *Reviews of Modern Physics* 73 (3): 681–700. http://link.aps.org/doi/10.1103/RevModPhys.73.681.

Heeger, A. J., and M. A. Diaz-Garcia. 1998. Semiconducting polymers as materials for photonic devices. *Current Opinion in Solid State and Materials Science* 3 (1): 16–22. http://www.sciencedirect.com/science/article/pii/S1359028698800600.

Henkel, K., A. Oprea, I. Paloumpa, G. Appel, D. Schmeißer, and P. Kamieth. 2001. Selective polypyrrole electrodes for quartz microbalances: NO2 and gas flux sensitivities. *Sensors and Actuators B: Chemical* 76 (1–3): 124–129.

Hosseini, S. H., and A. A. Entezami. 2003. Conducting polymer blends of polypyrrole with polyvinyl acetate, polystyrene, and polyvinyl chloride based toxic gas sensors. *J. Appl. Polym. Sci.* 90 (1): 49–62.

Hwang, J., J. Y. Yang, and C. W. Lin. 2001. Recognition of alcohol vapor molecules by simultaneous measurements of resistance changes on polypyrrole-based composite thin films and mass changes on a piezoelectric crystal. *Sensors and Actuators B: Chemical* 75 (1–2): 67–75.

Ivanov, P., E. Llobet, X. Vilanova, J. Brezmes, J. Hubalek, and X. Correig. 2004. Development of high sensitivity ethanol gas sensors based on Pt-doped SnO2 surfaces. *Sensors and Actuators B—Chemical* 99 (2–3): 201–206. http://www.sciencedirect.com/science/article/pii/S0925400503008359.

Janata, J., and M. Josowicz. 2003. Conducting polymers in electronic chemical sensors. *Nat Mater* 2, 1: 19–24.

Jin, Z., Y. Su, and Y. Duan. 2001. Development of a polyaniline-based optical ammonia sensor. *Sensors and Actuators B: Chemical* 72 (1): 75–79.

Josowicz, M., H. S. Li, K. Domansky, and D. R. Baer. 1999. Effect of oxidation state of palladium in polyaniline layers on sensitivity to hydrogen. *Electroanalysis* 11(10–11): 774–781.

Karin, P. K. 2001. Conducting polymer-based Schottky barrier and heterojunction diodes and their sensor application. In *Handbook of Surfaces and Interfaces of Materials*, ed. H. S. Nalwa, 445–494. Burlington: Academic Press.

Kawase, T., T. Shimoda, C. Newsome, H. Sirringhaus, and R. H. Friend. 2003. Inkjet printing of polymer thin film transistors. *Thin Solid Films* 438–439: 279–287. http://www.sciencedirect.com/science/article/pii/S0040609003008010.

Kirner, U., K. D. Schierbaum, W. Göpel, B. Leibold, N. Nicoloso, W. Weppner, D. Fischer, and W. F. Chu. 1990. Low and high temperature TiO2 oxygen sensors. *Sensors and Actuators B—Chemical* 1 (1–6): 103–107. http://www.sciencedirect.com/science/article/pii/092540059080181X.

Koll, A., S. Kawahito, F. Mayer, C. Hagleitner, D. Scheiwiller, O. Brand, and H. Baltes. 1998. A flip-chip-packaged CMOS chemical microsystem for detection of volatile organic compounds. *SPIE Proceeding Series* 3328: 223–232.

Kros, A., R. J. M. Nolte, and N. A. J. M. Sommerdijk. 2002. Poly(3,4-ethylenedioxythiophene)-based copolymers for biosensor applications. *J. Polym. Sci. A Polym. Chem.* 40 (6): 738–747.

Kurosawa, S., N. Kamo, D. Matsui, and Y. Kobatake. 1990. Gas sorption to plasma-polymerized copper phthalocyanine film formed on a piezoelectric crystal. *Analytical Chemistry* 62 (4): 353–359. http://dx.doi.org/10.1021/ac00203a009.

Lancet, D. 1991. The strong scent of success. *Nature* 351 (6324): 275–276. http://dx.doi.org/10.1038/351275a0.

Lange, U., and V. M. Mirsky. 2011. Chemiresistors based on conducting polymers: A review on measurement techniques. *Analytica Chimica Acta* 687(2): 105–113

Liu, Y., and T. Cui. 2005. Polymer-based rectifying diodes on a glass substrate fabricated by ink-jet printing. *Macromolecular Rapid Communications* 26 (4): 289–292. http://dx.doi.org/10.1002/marc.200400485.

Lundstrom, I., S. Shivaraman, C. Svensson, and L. Lundkvist. 1975. A hydrogen-sensitive MOS field-effect transistor. *Applied Physics Letters* 26 (2): 55–57.

MacDiarmid, A. G. 2001. Nobel lecture: Synthetic metals: A novel role for organic polymers. *Reviews of Modern Physics* 73 (3): 701–712. http://link.aps.org/doi/10.1103/RevModPhys.73.701.

MacDiarmid, G. and A. J. Epstein. 1994. The concept of secondary doping as applied to polyaniline. *Synthetic Metals* 65 (2–3): 103–116.

MacDiarmid, G., and A. J. Epstein. 1995. Secondary doping in polyaniline. *Synthetic Metals* 69 (1–3): 85–92.

McGehee, M. D., and A. J. Heeger. 2011. Semiconducting (conjugated) polymers as materials for solid-state lasers. *Advanced Materials* 12 (22): 1655–1668. http://dx.doi.org/10.1002/1521–4095(200011)12:22<1655::AID-ADMA1655>3.0.CO;2–2.

Milella, E., and M. Penza. 1998. SAW gas detection using Langmuir–Blodgett polypyrrole films. *Thin Solid Films* 327–329: 694–697.

Moraes, F., Y. W. Park, and A. J. Heeger. 1986. Soliton photogeneration in trans-polyacetylene: Light-induced electron spin resonance. *Synthetic Metals* 13 (1–3): 113–122. http://www.sciencedirect.com/science/article/pii/0379677986900615.

Morita, Y., K. I. Nakamura, and C. Kim. 1996. Langmuir analysis on hydrogen gas response of palladium-gate FET. *Sensors and Actuators B—Chemical* 33 (1–3): 96–99. http://www.sciencedirect.com/science/article/pii/0925400596019569.

Nanto, H., Y. Yokoi, T. Mukai, J. Fujioka, E. Kusano, A. Kinbara, and Y. Douguchi. 2000. Novel gas sensor using polymer-film-coated quartz resonator for environmental monitoring. *Materials Science and Engineering: C* 12 (1–2): 43–48. http://www.sciencedirect.com/science/article/pii/S0928493100001569.

Nelson, S. G., K. S. Johnston, and S. S. Yee. 1996. High sensitivity surface plasmon resonance sensor based on phase detection. *Sensors and Actuators B—Chemical* 35 (1–3):
187–191. http://www.sciencedirect.com/science/article/pii/S0925400597800524.

Nguyen Van, C., and K. Potje-Kamloth. 2001. Electrical and NOx gas sensing properties of
metallophthalocyanine-doped polypyrrole/silicon heterojunctions. *Thin Solid Films* 392
(1): 113–121.

Nylander, C., M. Armgrath, and I. Lundstrom. 1983. An ammonia detector based on a conducting polymer. In *Proceedings of the International Meeting on Chemical Sensors*,
203–207.

Patil, S., J. R. Mahajan, M. A. More, and P. P. Patil. 1999. Electrochemical synthesis of poly(o-
methoxyaniline) thin films: Effect of post treatment. *Materials Chemistry and Physics*
58 (1): 31–36.

Pearce, T. C., S. S. Schiffman, H. T. Nagle, and J. W. Gardner. 2003. *Handbook of machine
olfaction—Electronic nose technology*. New York: John Wiley & Sons.

Peet, J., J. Y. Kim, N. E. Coated, W. L. Ma, D. Moses, A. J. Heeger, and G. C. Bazan. 2007.
Efficiency enhancement in low-bandgap polymer solar cells by processing with alkane
dithiols. *Nature Materials* 6 (7): 497–500. http://dx.doi.org/10.1038/nmat1928.

Penza, M., E. Milella, and V. I. Anisimkin. 1998. Monitoring of NH_3 gas by LB polypyrrole-
based SAW sensor. *Sensors and Actuators B: Chemical* 47(1–3): 218–224.

Péres, L. O., R. W. C. Li, E. Y. Yamauchi, R. Lippi, and J. Gruber. 2012. Conductive polymer
gas sensor for quantitative detection of methanol in Brazilian sugar-cane spirit. *Food
Chemistry* 130 (4) 1105–1107.

Persaud, K., and G. Dodd. 1982. Analysis of discrimination mechanisms in the mammalian olfactory system using a model nose. *Nature* 299 (5881): 352–355. http://dx.doi.
org/10.1038/299352a0.

Persaud, K. C., and P. Pelosi. 1985. An approach to an artificial nose. *Transactions of the
American Society for Artificial Internal Organs* 31: 297–300.

Pron, A., and P. Rannou. 2002. Processible conjugated polymers: From organic semiconductors to organic metals and superconductors. *Progress in Polymer Science* 27 (1):
135–190.

Purbrick, J. A. 1981. A force transducer employing conductive silicon rubber. In *Proceedings
of the 1st International ROVISEC Conference*, 73–80.

Ressler, K. J., S. L. Sullivan, and L. B. Buck. 1994. Information coding in the olfactory system: Evidence for a stereotyped and highly organized epitope map in the olfactory bulb.
Cell 79 (7): 1245–1255. http://linkinghub.elsevier.com/retrieve/pii/0092867494900159.

Rivera, D., M. K. Alam, C. E. Davis, and C. K. Ho. 2003. Characterization of the ability
of polymeric chemiresistor arrays to quantitate trichloroethylene using partial least
squares (PLS): Effects of experimental design, humidity, and temperature. *Sensors and
Actuators B: Chemical* 92 (1–2): 110–120.

Rungnapa, T. 2003. Effect of nitrogen dioxide and temperature on the properties of lead phthalocyanine in polypyrrole. *Thin Solid Films* 438–439: 14–19.

Sakurai, Y., H. S. Jung, T. Shimanouchi et al. 2002. Novel array-type gas sensors using conducting polymers, and their performance for gas identification. *Sensors and Actuators
B: Chemical* 83 (1–3): 270–275.

Salaneck, W. R., and J. L. Brédas. 1994. Conjugated polymers. *Solid State Communications*
92 (1–2): 31–36. http://www.sciencedirect.com/science/article/pii/0038109894908559.

Santonico, M., P. Pittia, G. Pennazza, E. Martinelli, M. Bernabei, R. Paolesse, A. D'Amico,
D. Compagnone, and C. Di Natale. 2008. Study of the aroma of artificially flavoured cus-
tards by chemical sensor array fingerprinting. *Sensors and Actuators B—Chemical* 133
(1): 345–351. http://www.sciencedirect.com/science/article/pii/S0925400508001585.

Shepherd, G. M. 1987. A molecular vocabulary for olfaction. *Annals of the New York Academy
of Sciences* 510 (1): 98–103. http://dx.doi.org/10.1111/j.1749–6632.1987.tb43474.x.

Shepherd, G. M. 1994. Discrimination of molecular signals by the olfactory receptor neuron. *Neuron* 13 (4): 771–790. http://www.sciencedirect.com/science/article/pii/0896627394902453.

Shirakawa, H. 2001. Nobel lecture: The discovery of polyacetylene film: The dawning of an era of conducting polymers. *Reviews of Modern Physics* 73 (3): 713–718. http://link.aps.org/doi/10.1103/RevModPhys.73.713.

Shirakawa, H., E. J. Louis, A. G. MacDiarmid, C. K. Chiang, and A. J. Heeger. 1977. Synthesis of electrically conducting organic polymers: Halogen derivatives of poly-acetylene, (CH). *Journal of the Chemical Society Chemical Communications* (16): 578–580. http://dx.doi.org/10.1039/C39770000578.

Taylor, J. E., F. K. C. Harun, J. A. Covington, and J. W. Gardner. 2009. Applying convolution-based processing methods to a dual-channel, large array artificial olfactory mucosa. *AIP Conference Proceedings* 1137 (1): 181–184. http://link.aip.org/link/?APC/1137/181/1.

Vassar, R., S. K. Chao, R. Sitcheran, J. M. Nuñez, L. B. Vosshall, and R. Axel. 1994. Topographic organization of sensory projections to the olfactory bulb. *Cell* 79 (6): 981–991. http://linkinghub.elsevier.com/retrieve/pii/0092867494900299.

Virji, S., J. Huang, R. B. Kaner, and B. H. Weiller. 2004. Polyaniline nanofiber gas sensors: Examination of response mechanisms. *Nano Lett.* 4 (3): 491–496.

White, J., J. S. Kauer, T. A. Dickinson, and D. R. Walt. 1996. Rapid analyte recognition in a device based on optical sensors and the olfactory system. *Analytical Chemistry* 68 (13): 2191–2202. http://dx.doi.org/10.1021/ac9511197.

Winquist, F., S. Holmin, C. K. Rülcker, P. Wide, and I. Lundström. 2000. A hybrid elec-tronic tongue. *Analytica Chimica Acta* 406 (2): 147–157. http://www.sciencedirect.com/science/article/pii/S0003267099007679.

Xie, D., Y. Jiang, W. Pan, D. Li, Z. Wu, and Y. Li. 2002. Fabrication and characterization of polyaniline-based gas sensor by ultra-thin film technology. *Sensors and Actuators B: Chemical* 81 (2–3): 158–164.

Yokoi, M., K. Mori, and S. Nakanishi. 1995. Refinement of odor molecule tuning by dendro-dendritic synaptic inhibition in the olfactory bulb. *Proceedings of the National Academy of Sciences* 92 (8): 3371–3375. http://www.pnas.org/content/92/8/3371.abstract.

Zemel, J. N. 1996. Future directions for thermal information sensors. *Sensors and Actuators A: Physical* 56 (1–2): 57–62. http://www.sciencedirect.com/science/article/pii/0924424796012836.

Zhang, Q. T., P. Wang, J. P. Li, and X. G. Gao. 2000. Diagnosis of diabetes by image detection of breath using gas-sensitive laps. *Biosensors and Bioelectronics* 15 (5–6): 249–256. WOS: 000089512200004.

4 The Synthetic Moth: A Neuromorphic Approach toward Artificial Olfaction in Robots

Vasiliki Vouloutsi, Lucas L. Lopez-Serrano, Zenon Mathews, Alex Escuredo Chimeno, Andrey Ziyatdinov, Alexandre Perera i Lluna, Sergi Bermúdez i Badia, and Paul F. M. J. Verschure

CONTENTS

4.1 INTRODUCTION

Olfaction is a sense that is vital for many living organisms. Animals have been relying on smell to sample the environment and gather information from it. Olfaction enables the identification of food, mates, and predators as well as communication (Mykytowycz 1985) not only between members of the same or different species but also between animals and the environment.

Nevertheless, olfaction has not been as widely studied as vision or the auditory system. A deeper understanding of the biological olfactory system would allow us to develop novel artificial olfactory systems for real-world robotic applications such as environmental monitoring (Trincavelli et al. 2008), land mine detection (Bermudez i Badia et al. 2007), as well as detection of explosives and other hazardous substances (Rachkov et al. 2005; Distante et al. 2009). Although there have been several attempts to implement the sense of smell on robots, biological olfaction outperforms its artificial counterparts in robustness, size, response time, precision, and complexity. Animals, and more specifically insects with relatively simple nervous systems, are able to unravel the problem of odor localization and classification with great efficiency: bees use odor to localize nests, ants use pheromone trails to organize foraging in swarms, male moths use olfaction to locate mates (Baker and Haynes 1987), and so on.

Despite the technological advances in the field of artificial olfaction, a robust solution for the task of odor source localization and classification utilizing a fully autonomous robot has not yet been demonstrated. The main challenge is thus to develop an intelligent system able to robustly encode and decode odors as well as navigate autonomously in natural environments and successfully locate an odor source.

Artificial olfaction remains a challenging field in research, as it postulates the development of chemical sensors that are able to reliably capture information from the environment. In the field of artificial chemical sensing there is a wide diversity of technologies; however, the most widely used chemical sensors are made of thin-film metal oxide (MOX). These chemical sensors provide a broad spectrum of sensitivity to volatile chemical compounds with low power consumption. When employed on a robotic platform, they are usually structured in arrays of different types of chemical sensors—widely known as e-noses—which provide less error rates and a larger scale of chemical detection. Nevertheless, they are still less efficient than the sensory modalities of animals. As an alternative to an artificial chemical sensor Kuwana et al. (1999) have used its biological counterpart, which is the actual antennal lobe of a living silkworm moth connected to a mobile robot so as to perform pheromone search.

Equipping a robot with reliable chemical sensors is not enough to perform the odor classification and localization task. This task requires the development of robust odor classification models as well as odor source localization strategies that handle and exploit the information acquired from both the classification model and other sensory modalities. Early attempts to achieve the odor localization task are demonstrated by the Braitenberg's vehicles (Gomez-Marin et al. 2010; Lilienthal and Duckett 2003) or high-level processes that include a planner and symbolic reasoning (Loutfi and Coradeschi 2008). In the past two decades, several attempts have been made to model animals' behaviors and techniques to achieve a robust odor localization and classification system. For instance, to determine the direction of a gas source, Hiroshi Ishida and Atsushi Kohnotoh (2008) based their model on the dog's nose. Frank Grasso et al. (2009) have modeled the behavior of a lobster and built a robot that performs the odor localization task in an underwater environment. The list of studies that approach artificial olfaction by modeling animal olfaction is constantly increasing, with an emphasis on insect chemolocalization, and most specifically, the chemical search based on the behavior and neural substrates of the male moth (Pyk et al. 2006). In fact, in a comparative study of robot-based odor source localization strategies (Bermudez i Badia and Vershure 2009), the authors compare reactive approaches with strategies employed by the male moth, concluding that the latter are more efficient in correct localizations.

Nonetheless, to locate the source of a chemical compound in real-world applications is a rather difficult task. Odors are chemical volatiles in the atmosphere that are mainly transported by airflow, creating a plume. However, the plume dispersion dynamics vary greatly depending on the medium, as the interaction of the airflow with other surfaces produces turbulence. This dispersion is best described by the so-called Reynolds number. In fluid mechanics, the Reynolds number can be characterized by different conditions, where a fluid may be in relative motion to a surface. It includes density and viscosity and measures the ratio of inertial forces to viscous forces. With low Reynolds numbers where viscosity prevails, there is a smooth constant fluid motion with a monotonic decrease of the chemical concentration. At medium or high values, however, turbulence dominates, producing flow instabilities. To address the problem of odor localization and classification, Kowaldo has proposed to divide the task of odor localization in three general steps: (1) search for and identify the chemical compound of interest, (2) track the odor using several sensory modalities (such as chemical), and (3) identify the source of the odor (by either vision or olfaction). Consequently, different search and classification strategies need to be employed for different environments (Kowaldo and Russell 2008).

Our aim is to achieve a novel olfactory-based system that will allow an autonomous mobile robot to navigate within a given environment and locate the source of the desired odor. We propose two models for classification and localization based on the neural substrates and mechanisms employed by a biological system that is known to perform the task of odor localization and classification in a robust way—the male moth. To assess our models, we have conducted experiments using two different chemical compounds: ethanol and ammonia. Our results show the first steps toward a stable odor localization and classification system.

4.2 THE MOTH

4.2.1 Moth Behavior

Insects in general are particularly good at using chemical cues to analyze the environment and achieve key objectives such as locate food, find mates, or communicate with each other. In particular, moths have been widely studied due to their ability to locate the female moth from a large distance, up to several hundred meters. What moths are detecting as odor stimuli are specific pheromone blends. These pheromones are mixed in a complex chemical background and are diffused in turbulent plumes. However, male moths are able to detect minute concentrations of pheromone and locate the female that is emitting them. Thus, moths are able to solve the odor classification and localization task by combining active sampling with specific behavioral and information processing strategies.

The female moth releases a species-specific pheromone blend that acts as a sex attractant for the male moths. This blend flows downwind, creating a specific plume shape. The plume has a filamentous structure. Once the male moth detects the molecules of pheromone within the plume, it flies slowly upwind, tracing the filament of the plume. This stereotypical behavior is called surge (Pearce et al. 2004). However, due to the dynamics of the plume and the complexity of its structure, the moth often loses track of the pheromone plume during surging. To re-acquire the track of the plume, moths have developed cast behavior, which is basically a zigzag movement orthogonal to the wind direction (Pearce et al. 2004) (Figure 4.1). Interestingly enough, when the male moth loses track of the pheromone plume, and after casting finds it again, the point in which it has re-acquired the plume is usually closer to the source than when it initially lost it.

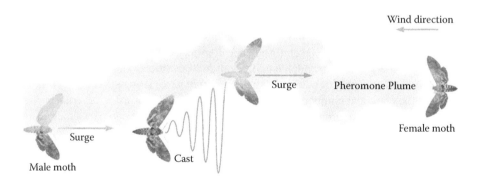

FIGURE 4.1 Illustration showing the pheromone plume and the male moth cast-and-surge behaviors. (Courtesy of Lopez, L. L. et al., in *On Biomimetics*, ed. L. D. Pramatarova, InTech, 2011. Accessed from http://www.intechopen.com/books/on-biomimetics/moth-like-chemo-source-localization-and-classification-on-an-indoor-autonomous-robot.)

As a result, the behavior employed by the male moth, when it tries to locate the female by tracking the pheromone plume, relies on complex information acquired by both anemotaxis (the orientation movement of the moth in response to the wind) and chemotaxis (the direction or movement of the moth according to a chemical compound).

Given this background and in order to understand the neural substrates and mechanisms that endow male moths with such robustness and high performance, we have developed a model that is based on behavioral mechanisms employed by the male moth, the so-called cast-and-surge behavior, and implemented the resulting model in an autonomous robot. By testing the behavior of our robot in real-world experiments, we want to be able to verify and strengthen our models, and therefore push forward our understanding of the mechanisms employed by the male moth.

4.2.2 OLFACTORY PATHWAY

The main components of the insect olfactory pathway are the olfactory receptor neurons (ORNs) in the antenna, the antennal lobe (AL), and the mushroom body (MB) (Hansson 2002) (Figure 4.2). The ORNs are located in the olfactory epithelium in the antenna and project their axons through the olfactory nerve to the insect antennal lobe. They respond to different chemical stimulus present in the air. The number of glomeruli is closely related to the number of types of ORNs. This organization is likely to help the AL to deal with noisy conditions and dynamic input (Laurent 1999). The glomeruli signals are sent to two different types of neurons: projection neurons (PNs) and local neurons (LNs). The projection neurons are the output of the AL to the MB, and will spike simultaneously in the presence of a specific odor. LNs laterally interconnect with the PNs and modify their activity by means of inhibition. Finally, the MB is responsible for the odor's memory and learning processing.

4.2.2.1 Olfactory Receptor Neurons

The function of the ORNs is to send precise information to the nervous system on the amount of single odorants present in the air. When an odorant enters in contact with an ORN it interacts with the receptor proteins in the membrane of the neuron and increases the membrane potential, eventually generating a spike. The spike amplitude may differ from ORN to ORN, but it is thought not to carry any useful information (Todd and Baker 1999). What is definitely important is the frequency of the spikes, and also the temporal pattern they create. This translates into a constantly firing rate going from the ORN to the AL that indicates the odorant concentration. Although different ORNs respond differently to different odorants, such differences are in some cases very small and not easy to observe, which makes odor classification a nontrivial task. ORNs are present all over the antenna, also providing spatial information on where the odorant is located in the environment. This information is of fundamental importance for the flying strategy of the insect when locating a plume. The distribution of the different types of ORNs varies in different species, but generally they will be found homogeneously distributed along the antenna.

A key factor for classification is the response of the ORNs over time to a constant stimulus. As to be expected, ORNs do not react immediately to an odorant, but for any given concentration of the odor a certain time is needed to reach the maximum

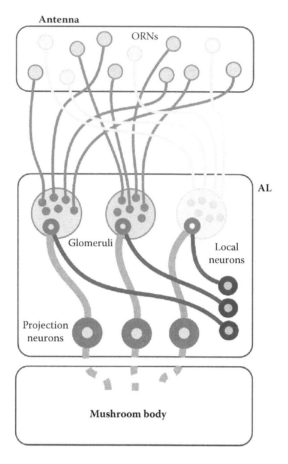

FIGURE 4.2 Functional representation of the insect's olfactory pathway. ORNs belonging to the same class converge onto the same glomerulus. LNs interconnect with the PNs, which are then connected to higher brain areas such as the MB. (Courtesy of Lopez, L. L. et al., in *On Biomimetics*, ed. L. D. Pramatarova, InTech, 2011. Accessed from http://www.intechopen.com/books/on-biomimetics/moth-like-chemo-source-localization-and-classification-on-an-indoor-autonomous-robot.)

firing rate, and much longer time is needed to reduce their activity when the odor is removed (Purves et al. 2001). Curiously, this behavior has been shown to be very useful for odor blend identifications, since those times are different for different particles.

4.2.2.2 Antennal Lobe

In most species the AL has the shape of a sphere and is well demarcated from other parts of the brain. It is composed by a number of neuropilar compartments called glomeruli. Glomeruli are spheroidal structures that compile the activity that comes from ORNs of the same type, although in some cases different kinds of ORNs are mixed. Moreover, the male moth has a pheromone-specific macro-glomerulus for reproductive purposes.

There are basically two types of neurons in the AL: the projection neurons (PNs) that filter the activity from the ORNs and then send a processed signal to higher cognitive areas, and the local neurons (LNs) that shape this activity by extracting the most significant characteristics of the signal. The LNs can be at the same time of two types: heterogeneous LNs (local neurons whose inputs and outputs remain in the same glomerulus) or homogeneous (local neurons whose inputs and outputs interact with different glomeruli).

The AL shows a high level of interconnectivity between LNs and PNs. The number of LNs is at least four times that of the PNs. A LN may receive input from one or more glomeruli, and may inhibit another LN or a PN. This inhibition depends on two different types of GABA receptor: GABAA (fast) and GABAB (slow) (Waldrop et al. 1987). It is still not clear if connections going from the glomeruli to the PNs exist. It is believed that the inhibition arising from LNs modifies the activity of the ORNs and shapes the response of PNs, which project to the MB (Christensen et al. 1993).

4.2.2.3 Mushroom Body

The mushroom body (MB) is known to be involved with the learning and memory of odors. This work is to focus on the role of the AL in the signal processing for the classification task. The role of the MB is secondary and will be considered just as a linear classifier.

4.3 METHODS

This section describes the technical characteristics of the robotic platform and sensors employed for this study, as well as the technical specifications of the embedded computer and the software used. We also provide an outline of our experimental setup and the evaluation tasks we applied to test our system.

4.3.1 THE ROBOT

The autonomous robot described here was developed within the European project NEUROCHEM supported by Bio-ICT and was the product of collaboration between UPF (Universitat Pompeu Fabra) and UPC (Universitat Politecnica de Catalunya). The robot is composed of two main parts: the mobile robotic platform developed in SPECS-UPF and an embedded computer assembled at UPC. The basic requirements applied to the robot include a full-functioning interface with chemical and other sensors, full autonomy, and demonstration capabilities.

4.3.1.1 Sensory Modalities—Hardware

We provided our robot with several sensory modalities, so as to be able to navigate freely and explore the environment. To receive information on the distance of objects on the left, middle, and right part of the robot, three SRF08 ultrasonic sensors (Devantech Ltd. {Robot Electronics], Norfolk, England) were placed, equally spaced, in the front part of the robotic platform. We also used a CMPS03 compass (Devantech Ltd. {Robot Electronics], Norfolk, England), especially designed for

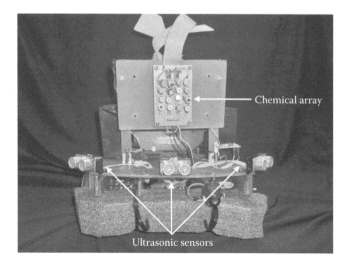

FIGURE 4.3 Picture of the robotic platform. The chemosensor board is placed in the middle of the platform and the ultrasonic sensors are placed in the front part of the robot in equal places.

robotic navigation, as it produces a unique number to represent the robot direction. The wind direction was measured with a custom-built wind vane that also produces a unique number for the direction of the wind, relative to the direction of the robot. Finally, we equipped the robot with an array of 16 MOX chemical sensors. The board is placed in the middle of the platform to avoid the sensors being affected by rotations (Figure 4.3).

Additionally, a GPS and a two-axis accelerometer were applied to the robotic platform, but were not utilized in the current set of experiments. The moving platform is based on an Arduino board with a bluetooth interface. The mobile base is interconnected with the embedded platform via a bluetooth dongle. The array of the chemical sensors is directly connected to the embedded computer, allowing us to acquire real-time data from the environment. In addition to that, a wireless LAN adapter has been utilized allowing connections between the embedded computer and a local network created for the purpose of our experiments. This ensures communication with the embedded and any other computer connected to the network.

4.3.1.2 Chemical Array

The success of the odor classification and localization task highly depends on the instrumentation capabilities of the robot for odor sensing. The robot's design is able to host three types of gas sensor arrays. The first type is a large-scale array of 64K polymeric sensors (Beccherelli et al. 2009), consisting of 16 modules of 64 × 64 elements each and approximately 8 sensor types. The second sensor array is composed of four types of thin-film metal oxide (MOX) Figaro sensors (Figaro USA, Inc., Arlington Heights, IL, United States). All four types of Figaro sensors have a low power consumption, small size, and long life expectancy. The four types of MOX

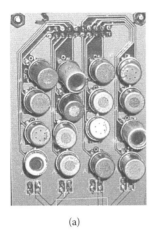

TGS 2612	TGS 2442	TGS 2600	TGS 2610
TGS 2610	TGS 2610	TGS 2610	TGS 2610
TGS 2600	TGS 2612	TGS 2600	TGS 2442
TGS 2442	TGS 2610	TGS 2610	TGS 2612

(a)   (b)

FIGURE 4.4   (a) The board of the 4×4 chemical sensor array. (b) The spatial arrangement of the four types of Figaro chemical sensors: TGS 2442, TGS 2600, TGS 2610, and TGS 2612.

sensors are TGS 2442, TGS 2600, TGS 2610, and TGS 2612. Each one of them is a broadly selective gas sensor. Figure 4.4 shows the spatial arrangement of each sensor. Finally, the third type of gas sensor array is a virtual sensor array, which is basically a software abstraction of sensor signals that were used to test various models of insect olfaction. The results represented in this work are obtained with the second type of MOX sensor array.

4.3.1.3   Robotic Platform

A first version of the robotic platform used in our experiments had a tank structure with two motors: one in the front and one in the back, and caterpillar tracks on each side (Figure 4.5). With this platform, the robot operated at 80% of nominal speed, advancing 1.8 m/s. However, this given speed was considered inconsistent with the response time of the sensors, and it was necessary to reduce the speed. Ideally, the robot should be able to move at a speed of 3 cm/s. Nevertheless, due to the nature of the motors, it was not possible to lower the speed significantly, as they required a minimum of 60% of tension to begin to move. After having optimized the robot with the minimum possible speed, we tested the platform inside the wind tunnel, by performing one simple cast. Although this design was favoring movement through different terrains and supported the weight of the embedded device as well as the batteries, it did not allow controlled movements and slow maneuvers, which were considered necessary for chemo-search inside the wind tunnel.

Therefore, we decided to redesign the robot to improve maneuverability in relation to speed and ability to carry the weight of the embedded computer and its batteries. The new robot differed from the previous one in the design of the platform, as this one was equipped with two wheels instead of caterpillars. The rest of the sensors that were employed for the chemo-search were mounted onto the new one. The design of the new robot was based on the structure of the

FIGURE 4.5   Image of the autonomous robotic platform when operating with caterpillar tracks. The embedded computer, batteries, and sensory modalities, as well as the chemo-sensor array, are placed on top of the robotic platform. (Courtesy of Lopez, L. L. et al., in *On Biomimetics*, ed. L. D. Pramatarova, InTech, 2011. Accessed from http://www.intechopen. com/books/on-biomimetics/moth-like-chemo-source-localization-and-classification-on-an-indoor-autonomous-robot.)

tractor-carrying aircrafts, due to its maneuverability. The structure consisted of two wheels, one on each side, controlled by one motor each independently (Figure 4.6). To allow fine-tuning of the robot, we have placed at the back of the platform a set of three omnidirectional passive wheels, and the load of the robot (embedded, batteries) is placed in the front. To reduce the speed of the robot further, we have applied a reduction gear system on each motor. Just like the previous robot, we have tested this new structure by performing a simple cast task. The results show that the novel platform has positive effects on both maneuverability and performance.

4.3.1.4   Embedded System

Research in biomimetic algorithms on artificial olfaction poses new technical requirements to the hardware and software equipment of the sniffing robots. The embedded technology implemented in the synthetic moth robot offers several benefits in this area. The modular structure of the embedded platform assigns to each part the involved tasks, including system control, data acquisition, biologically inspired processing, and visualization. The system runs a GNU/Linux image that can be operated either headless or with the aid of the standard graphical user interface (GUI) solution, with the iqr simulator for large-scale neural systems embedded in the software (Bernardet and Verschure 2010).

The architecture of the embedded computer is based on the PC104 standard, which is typically targeted to the industrial rugged embedded applications, where this technology permits data acquisition on extreme environments. The PC104 bus

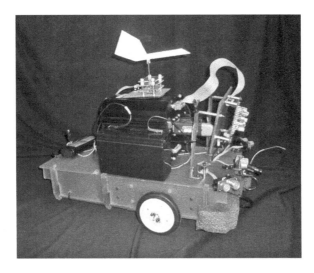

FIGURE 4.6 Image showing the robotic platform after the caterpillar tracks were replaced with wheels. This new design allowed better maneuverability of the robot.

offers additional benefits in terms of compact form factor (size reduces to 3.6 by 3.8 in.), a low number of components and internal connectors, and a low power consumption (1–2 W per module).

The embedded computer is composed of four PC104 component boards: CPU board PCM-3372F-S0A1E (Advantech), data acquisition board PC104-DAS16Jr/16 (Measurement Computing), power supply unit HESC104, and battery pack BAT-NiMh45 (Tri-m Systems). The CPU board is a single-board computer that provides a performance similar to a small laptop computer. The board contains an Intel VIA Eden V4 1.0 GHz processor, 1 GB RAM of DDR2 standard at 533 MHz, and the system chipset VIA CX700 with 64 MB VRAM. Running the models in the iqr neuronal simulator achieves 50 cycles per second, which was the target speed on the design of the system.

The I/O periphery of the CPU board consists of two serial ports, six USB 2.0, keyboard/mouse slots, audio and 8-bit GPIO ports, a 10/100 Mbps Ethernet interface, and a slot for a flash type I card. The data acquisition unit is a 16-channel board with 16-bit analog-to-digital converters (ADCs). The system is configured for a parallel 16-channel ADC at 100 KHz sampling frequency. Such a configuration is able to interface all the polymer gas sensor boards developed within the NEUROCHEM project, described in Section 4.3.2.2.

The power supply unit is a DC-DC converter with a wide range of input voltages, from 6 V to 40 V DC, and an output power of 60 W. The uninterruptible power supply (UPS) mode is supported with board configuration stored in the power board EEPROM memory. The power consumption of the embedded computer in the complete configuration is typically 9 W (maximum of 15.5 W). The NEUROCHEM polymeric sensor array with 64K elements with associated electronics requires from 4 W to 10 W. Given the maximum power consumption of 25.5 W, the system includes a battery pack of 4500 mA/h that guarantees an autonomous operation for around 1.1 h.

The software includes a software emulator of a large-scale sensor array. This software module permits us to work, test, benchmark, and prototype a complete neuromorphic signal processing tool chain without the requirement of a physical sensor array. The module includes means for the design of the experiment, and the generation of a large number of sensors/receptor, which behave realistically as a large array of polymeric sensors.

The virtualization of the hardware system has been also implemented in a custom GNU/Linux image based on the Debian (Live Debian project) operating system (OS). The released OS image includes drivers for the PC104 boards, custom data acquisition software, iqr modules of the developed neuromorphic models, and the model of chemosensory array. The end users can therefore choose to develop models targeting a physical sensor array mounted on the robotic platform, or test these models under a simulated experiment on a desktop computer without need of any specific hardware.

4.3.2 Software

4.3.2.1 iqr

Our system consists of two main models: classification and localization. In order to design and simulate the neural networks of both models, a solid software framework was needed. The tool we used is the open-source large-scale neural network simulator iqr (Bernardet and Verchure 2010) (available under the GPL license). iqr provides a multilevel neuronal simulation environment allowing us to visualize and analyze data in real time, as it supports interfacing to external devices, like robots, due to its modular structure. A great feature of iqr is that it is able to simulate biological nervous systems by using standard neural models, such as Linear Threshold or Integrate and Fire. In this way, all behaviors elicited originated from inhibitory and excitatory interactions among such neurons. Given our system's needs, specific modules allowing communication between the robot and different computers in the network have been developed using C++, as well as custom neurons. By employing and implementing our system with iqr, we were able to acquire data from the sensors of the robot, process them using the models of localization and classification, and send the output commands to the robot, in real time.

4.3.2.2 Data Acquisition Software

The main purpose of the acquisition software is to deliver the stream of real-time chemosensory signals to the classification and localization models implemented in the iqr framework. Figure 4.7 shows a diagram of the acquisition flow that consists of three levels: hardware, software, and user level. The neuromorphic models of the moth robot are located at the user level and interconnect with several components on the other two levels via the iqr modules. The chemosensory readings end up in the iqr modules passing though several stages at the software level. The low-level data acquisition is partly controlled by the Comedi-based driver.

The Comedi project develops open-source drivers, tools, and libraries for data open-source acquisition (Schleef et al. 2003). This project provides a collection of drivers for a variety of common data acquisition boards. The drivers are implemented

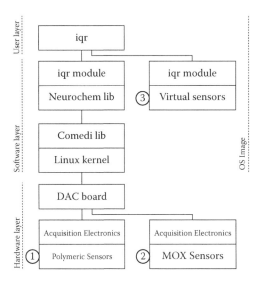

FIGURE 4.7 A scheme of the acquisition flow from three chemosensor arrays of different types: polymeric sensor array of up to 64K elements designed in the NEUROCHEM project, commercial metal oxide (MOX) sensors, and virtual sensors allowing generation of synthetic data (R package chemosensors).

with help of a Linux kernel module, offering common functionality and individual low-level driver tools. The functions are accessed via the Comedi user interface library, Comedilib. Available functions perform asynchronous and triggered acquisition, configuration of analog and digital channels, and the Direct Memory Access (DMA) data transfer to memory. The NEUROCHEM library is a shared library that communicates with the acquisition electronics of the sensor arrays. This library implements a custom signal protocol targeting the compatible sensor arrays by means of a developer-friendly interface wrapping Comedilib.

The NEUROCHEM acquisition software supports three types of arrays. The first array is a polymeric sensor array developed by the CNR Institute for Microelectronics and Microsystems, Rome, Italy, and the University of Manchester, UK (Beccherelli et al. 2009). This array is the first very large sensor array available, providing 64K sensing elements distributed between 16 sensor dies. Each sensor die has a size of 4 cm², and contains 4096 active sensor surfaces. Thirty-one different polymer types have been used on the deposition of the 64K sensing elements. This array is the most demanding stage for the software system. However, the slow dynamics of the chemical reactions in the sensor device limits the required acquisition speed, which is close to 1 s for a complete scan of all available sensors (64K).

The second array is a general purpose array made of 16 commercial metal oxide (MOX) sensors from Figaro Engineering, Inc., which has been developed at the University of Barcelona, Spain. This array is composed of 16 sensors of four different types (TGS 2442, TGS 2600, TGS 2610, and TGS 2612).

Both arrays, polymeric and metal oxide, are compatible with the same data acquisition protocol implemented in the NEUROCHEM driver, which takes care of each detail on the ADC process, including acquisition control, communications with the electronics in each sensor array, and signal filtering. The two arrays are available via the same iqr module designed for the end user.

As introduced in Section 4.3.1.4, the third array is a built-in software-based abstraction of a real polymer sensor array. This virtual sensor array is in fact exposing the functionality of an R package (R Development Core Team 2011) named chemosensors, developed by A. Ziyatdinov and A. Perera (R package chemosensors). This virtual array allows for the design of synthetic experiments that simulate real-time signals that have been used to test neuromorphic models within iqr on the NEUROCHEM project. The virtual array allows control of the generation of chemosensory stimuli with a variety of characteristics: unlimited number of sensors, support of multicomponent gas mixtures, and full parametric control of the noise in the sensors, including drift and nonlinearity. The R package chemosensors are included in the OS image released for the moth robot in the NEUROCHEM project.

4.3.3 Environmental Setup

4.3.3.1 Wind Tunnel

For the needs of our experimental setup, we have constructed a wind tunnel inside which the robot is placed. The wind tunnel is located at the SPECS lab in Barcelona, Spain. It consists of a wooden skeleton covered with a transparent polyethylene sheet of low density (Figure 4.8). This solution allows us to have a controlled indoor environment where the robot can move freely. A constant airflow is generated by four ventilators that are located at the one end of the wind tunnel. Each ventilator is

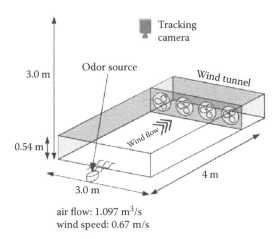

air flow: 1.097 m³/s
wind speed: 0.67 m/s

FIGURE 4.8 Layout of the wind tunnel. The camera is located 3 m above the wind tunnel. The arrows indicate the flow direction from the odor source to the exhaust ventilators. (Courtesy of Lopez, L. L. et al., in *On Biomimetics*, ed. L. D. Pramatarova, InTech, 2011. Accessed from http://www.intechopen.com/books/on-biomimetics/moth-like-chemo-source-localization-and-classification-on-an-indoor-autonomous-robot.)

a centrifugal 4.4 W fan that creates a negative pressure of an airflow velocity up to 1.0 ms^{-1}.

An odor source is placed on the upwind end of the tunnel. Therefore, the plume that is created moves across the whole wind tunnel from the point of the odor source to the four ventilators where the air is extracted out of the room. The wind tunnel is approximately 4 m long, 3 m wide, and 54 cm high.

For the scope of the odor localization experiment, the robot was placed in the middle of the wind tunnel, in front of the fans, facing upwind. In order to create the odor maps, the robot was placed in different parts of the wind tunnel in equal distances, facing upwind.

4.3.3.2   Vision-Based Tracking System (AnTS)

The trajectory of the robot was acquired with the general purpose video-based tracking system AnTS. The tracking system consists of a monochrome camera that is placed approximately 3 m above the wind tunnel. To track the robot independently of light conditions, an IR filter was applied to the camera and 3 IR LEDs were placed on the robot so that they could be identified by the camera. The AnTS tracking application is able to record not only the orientation and absolute position of the robot inside the wind tunnel, but also one trace per tracked element.

4.3.4   SYSTEM ARCHITECTURE

The developed models of odor localization and odor classification are based on the behavior and neural substrates of the male moth. The model of classification consists of three main stages, just like the olfactory pathway of the moth. The first stage is the ORN model, which groups the input from the sensors. The second stage is the AL model with custom modifications, which represents the stimulus information in such a way that is relatively easy to classify. Finally, a MB model acts as a linear classifier and obtains the identity of the blend. This information is then passed to the localization model, which is based on the two basic behaviors observed in the male moth: upwind movement (surge) and crosswind movement (cast). Our system combines active sampling from the environment with the moth's search strategy. This means that the system receives and processes in real time information acquired from the environment (such as odors, obstacles, etc.), decides which action to take, and then outputs the desired action to the motors (Figure 4.9).

4.3.4.1   ORN Model

The ORN model used is, in accordance with its analog in nature, taking the input from the chemical sensors and translating it to neural activity. The four different types of sensors the robot has are grouped in four different glomeruli, whose activity varies from 0 to 1 (Figure 4.10). There are, however, some important differences with the biological approach.

The presence of a stimulus is not represented with the firing rate of a neuron, but with its action potential. This way of realization is due to computational constraints in the embedded robot. Making it so, the neuron will show the spike activity on a specific time and not the time of each spike, saving a considerable number of cycles

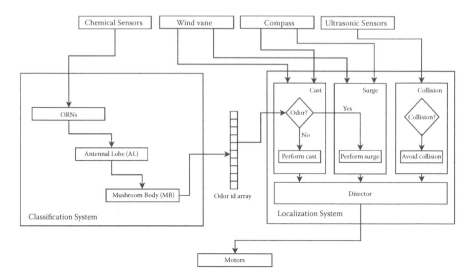

FIGURE 4.9 Overall system architecture. Inputs of the robot are represented on the top, while actuators (motors) are at the bottom of the picture. The classification system receives the input of the chemical sensors, which is grouped and normalized in the Olfactory Receptor Neurons (ORNs) group. It is later passed to the antennal lobe (AL), where the TPC process is performed. Finally, the mushroom body (MB) identifies the odor and translates it to an odor array, which acts as a link between both systems. According to whether there is an odor recognized or not, the localization system will decide if the action to perform is a cast or a surge, using the input of the wind vane and the compass (desired versus actual orientation). In parallel, the collision system detects any obstacle in the path and sends the order of avoidance if necessary. In the last step, the Director group decides which action is the one with the highest preference (collision avoidance, surge/cast).

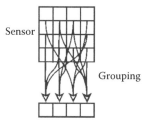

FIGURE 4.10 ORN model. First, similar sensor input represented by action potential is grouped into different glomeruli. Second, the signal is normalized to give values between 0 and 1.

by reducing the simulation speed. Second, the biological ORNs fire at a constant rate and decrease only in the presence of an odorant. However, even if this behavior is also given in the nature of the sensors, the output of the ORNs has been normalized to give a value between 0 and 1. Thus, the absence of odorant will be represented with a 0, and the saturation of the sensor given a concentration of particles will be 1.

Finally, each glomerulus in the AL receives input only from one type of sensor. In order to simplify the system, no sensor-mixed glomeruli were implemented.

4.3.4.2 Antennal Lobe Model

The classification model used in the robot is an adaptation of the one proposed by Knüssel (2006). This model is based in the so-called temporal population code (TPC).

After measuring the activity of the PNs in the moth with physiological methods, Knüssel proposed a theoretical model of the AL that takes into account not only the immediate firing rate value of a neuron, but also the evolution in time of the trends of spikes. As the main interest of Knüssel was to study temporal dynamics, his model is set to receive static sensory input. This model consists of four types of neurons: olfactory receptor neurons (ORNs), heterogeneous local neurons, homogeneous local neurons, and projection neurons. Except for the homogeneous local neurons, which are shared by all glomeruli, the rest of them can be found in the relation of one per glomerulus. Each glomerulus receives input from only one olfactory receptor neuron, which represents the average activity of all the receptor neurons of the same type. Glomeruli are physically arranged in a ring so they avoid the boundaries in the connections. The input from the olfactory receptor neuron excites the projection neuron, the homogeneous local neuron in the same glomeruli, and the homogeneous local neuron. It also provides excitation to the neighboring glomeruli heterogeneous local neurons. The homogeneous local neuron inhibits every heterogeneous local neuron with a fast synapse. This way, the homogeneous local neuron keeps the average firing rate of all the olfactory receptor neurons, while the heterogeneous local neurons in each glomerulus represent the difference of the glomeruli receptor neuron firing rate (or neighborhood of them) with the average. Projection neurons are at the same time inhibited with a slow synapse by the heterogeneous local neuron in the same glomeruli. This slow synapse corresponds to a standard exponential kernel, which increases the inhibition exponentially over time.

In this configuration, the model will react with a high, fast peak in the presence of a static stimulus in the receptor neuron, followed by a slow decrease, generating a so-called alpha function (Figure 4.11).

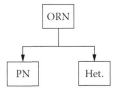

FIGURE 4.11 Illustration of the AL model proposed by Knüssel. Constant input from ORNs is passed to the PN in the glomerulus, the heterogeneous LN in the glomerulus and the neighboring ones, and a common homogeneous LN. Heterogeneous LNs in each glomeruli are inhibited by a fast-type connection from the homogeneous LN. At the same time, this heterogeneous LN inhibits its corresponding PN with an exponentially delayed connection, generating an alpha function in the PN if a constant stimulus is shown in the ORN for a short time.

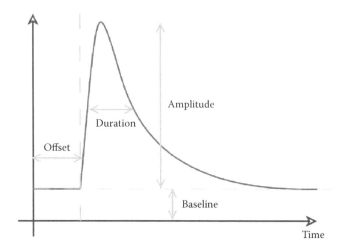

FIGURE 4.12   The alpha function is characterized by four parameters: baseline B, offset O, amplitude A, and duration D. According to Knüssel's model, if considering not only the amplitude but also the duration of this function, classification can be improved up to 40%.

This function encodes information about the stimuli in both amplitude and duration dimensions. While the initial amplitude (the peak) is defined by the direct excitation from the receptor neuron, the decay (and thus the duration) will be conditioned by the slow inhibition of the heterogeneous local neuron (meaning the relative position of the glomerulus receptor neuron with respect to the average of all of them). This way, both dimensions are critically important to the odor classification, being biologically compatible with the physiological measurements of the real moth projection neurons taken in (Knüssel et al. 2007). In this paper the alpha function (Figure 4.12) was used as a fit to the projection neurons' firing rate, and not only the amplitude but also the duration was shown to have important information content, improving the classification up to a 40% in comparison with simple spatial coding.

The model presented by Knüssel was implemented and adapted in order to make it viable to run in the embedded computer, but conserving the concept of the temporal population code (Figure 4.13).

The original model is prepared to respond to a constant, static input. Since Knüssel's objective was to show a dynamic response of the PNs to a static input from the ORNs, this was an ideal situation for his purpose. However, in this research a model able to receive dynamic and turbulent input is needed. The robot does not receive air-puffs in the sensors, but is moving inside the arena coping with turbulences.

By observing the reaction of the sensors in the wind tunnel, one realizes that their function over time already takes the shape of the previously described alpha function. The idea of this adaptation is to extract these noisy alpha functions when they are significant enough, clean them, and pass them to the MB for them to be identified.

To achieve this, a conditioning network was prepared for each glomerulus. The typical basic conditioning chain is described in Figure 4.14, although it can be made more complex by combining signals from several glomeruli and mixing them in different ways. On a first stage, the positive derivative of the signal coming from the

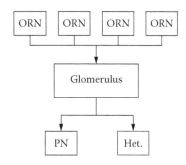

FIGURE 4.13 TPC model adaptations on Knüssel's model. Glomeruli are not interconnected anymore and the firing rate is represented by the neuron's activity.

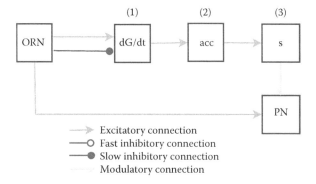

FIGURE 4.14 TPC model conditioning chain. First, the derivative is obtained via an immediate excitatory connection and a slow inhibitory connection (1). The accumulator group (2) keeps the signal of the derivative for some time. The S group (3) obtains a scale factor for the PN in function of the change of the ORN. PN is finally excited by the ORN and modulated by S.

glomerulus is obtained by feeding a target neuron with the immediate signal, and inhibiting it with a delayed signal. This allows the system to recognize whether the signal changes and how much, and to know if there is a significant alpha function in the entry. Second, this neuron excites another neuron, which has both a threshold and a membrane persistence value that keeps in memory for some milliseconds if there was a significant alpha in the entry, acting as an accumulator for the derivative. However, this last neuron is limited to never have a membrane potential higher than its input, so the membrane persistence is only used to gradually decrease the activity of the neuron through time. Subsequently, this value is normalized between 0 and 1 in the next layer, setting the boundaries by observing the common limits the sensors reached in the experiments. Doing so allows the use of the activity of this last layer to modulate the input from the glomeruli (also normalized previously), making the noise of most alpha functions disappear and reacting only when a significant change is produced on their input.

Another point is that the original model considers the firing rate as the main factor to be calculated. Although this is a more realistic approach, the run of all the spikes in the computationally limited system in the robot would require a notably high simulation speed, besides the transformation of the sensor input, which is of a continuous nature, also to spikes. To avoid this and save considerable computational power, the firing rate of each neuron is represented in our model by the activity of a linear threshold type of neuron, instead of using integrate and fire neurons, which are necessary for the spiking model. Implications of this are, however, minimal; the precision of the spikes is not critical for the shaping of the alpha function, and both amplitude and duration can be accurately shown.

Finally, in the simulations of the original model, 10 glomeruli are used in a ring, with each receptor neuron providing input not only to the corresponding glomerulus, but also to the one in the left and the one in the right. Since the robot has only four glomeruli, if we followed this strategy in our model, the relative position with respect to the average of each glomerulus represented by the heterogeneous local neuron would be almost insignificant and very sensitive to noise. Because of this, each receptor neuron in our model excites only the projection and local neurons existing on its own glomeruli.

4.3.4.3   Mushroom Body Model

The MB model that is used to classify the output of the AL is the one developed within the NEUROCHEM framework at University of Barcelona. It is a linear classifier that takes as input a group of neurons, some active, some not, and approximates them to the nearest activity pattern for which it was trained. In other words, it cleans the input and, if there is something similar to the collection of patterns it has inside, outputs the matching one.

The model is implemented as a module for iqr in C++ and has been ported to the robot in order to work with the input of both models of AL. The output of the TPC model of the antennal lobe needs to be adapted to be transformed from amplitude and time to a pure spatial spiking model. This is done by translating the response of every single projection neuron into a special array of neurons that retains the activity of the group for a specified time. This way, each of these groups representing a PN acts like a barcode characteristic of the odor blend. The task of the mushroom body is then to approximate this trend of pulses to the ones it was trained for before.

4.3.4.4   Localization Model

The model of localization is based on the behavioral strategies employed by the male moth when it is trying to locate the female. This model is responsible for receiving information sent by the classification model as well as the different sensory modalities of the robotic platform. Based on the information the system receives, it decides upon which action to take (such as avoid collision, cast, or surge). As this model talks directly to the motors, the behavior of the robot mainly relies on the motor's actions.

In Figure 4.15 we can see the localization model developed using iqr. The models in iqr are organized in two different levels: the top level represents the system and contains several processes, and in the process level each model is divided into units that allow us to interact with external devices. The main units of each process are groups of neurons that interact with each other through inhibitory and excitatory connections. As iqr is a large-scale neural simulator, each group represents a group

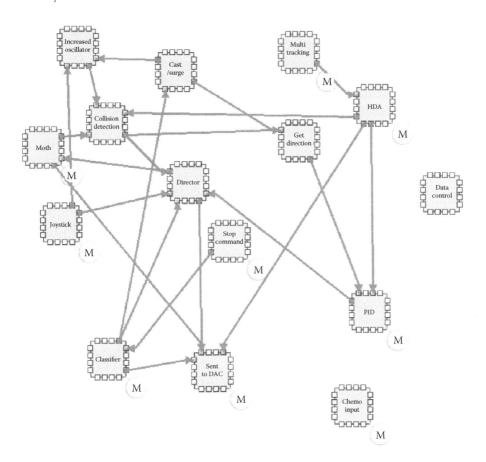

FIGURE 4.15 Image of the main system implemented in iqr. Each box represents a process. Each box with an M marking represents a module. The arrows indicate exchange of information between processes.

of similar neurons and each connection represents synapses of the same type that connect these groups, and each box represents a process. Each process that is marked with an M represents a module, developed in C++, that allows us to exchange data with an external device.

The main module that is responsible for the communication between the robotic platform and iqr is called Moth. This module receives information from the sensors (compass, wind vane, etc.) and makes them available to all other processes as well as outputs the desired commands to the robot's motors. The process that gets information from the ultrasonic sensors and decides whether there is a collision or not is the collision detection process. Depending on the readings of the sensors, it decides the appropriate desired direction for the motors so as to avoid collisions. To get the robot's position inside the testing arena, multitracking gets the x, y coordinates of the three points of the robot and transmits them to iqr, while HDA outputs the heading direction of the robot. PID gets as inputs the desired direction of the robot and the current direction and decides the corresponding movement of the motors (left, right, forward, backward) to get to the desired direction. The process that receives the information from the classification model is the classifier. It basically receives information regarding the odors the classifier has detected and informs the cast-and-surge process. It also receives a stop command, which it outputs to the motors, so as to allow the classifier some more time to detect a smell. Cast and surge is the process that decides, based on the information received by the classifier, whether a detected odor is the desired one, and thus performs cast (if there is no odor detected or not the desired one) or surge (if the target odor is detected by the classifier). Finally, one of the most important processes of the system is the director, as it sets the priority of each process over the motors.

4.3.4.4.1   Director

When two or more processes that run in parallel independently output commands to the same group of neurons (motors), there is a chance that two or more neurons will output a command to the motors at the same time. To avoid bewilderment, we have assigned a special process in order to avoid conflicts between processes that control the robot's motors. Therefore, only the process with the highest priority will finally send commands to the motors.

The architecture of the process is displayed in Figure 4.16. The neuronal group Final Motor Output is the group that commands directly the motors of the robot. In our system, the highest priority over all others (including collision avoidance or cast and surge) is the joystick, where a human controls the robot's movements. The second-in-hierarchy process is the stop command sent by the classifier, then the collision detection, followed by the PID.

4.3.4.4.2   Collision Detection

One of the most important processes of the system is that of collision detection. It receives as input the readings of the ultrasonic sensors and checks if there is a collision or not. The desired action (turn right or left) is based on both the sensors' readings and the current direction of the robot.

Figure 4.17 illustrates the iqr scheme of the collision detection process. The ultrasonic sensors measure the distance between the robot and an obstacle; thus, the readings may vary from 0 to 60 cm. We normalize these values with sensors 0–1 and set a threshold above which a collision will be detected. For that we need to reverse those values (where 0 would mean no object in the surrounding area), and the neuronal group Collision contains the reversed values of sensors 0–1.

If there is no collision, the neural group Decision Compass is inhibited; thus, there is no command outputted to the motors. However, if there is a collision, based on the robot's direction, the system decides whether to turn left or right and outputs the corresponding command to the motors.

4.3.4.4.3   Classifier

The process classifier is the bridge between the models of localization and classification. It is important to be able to detect that something is an odor and classify it and

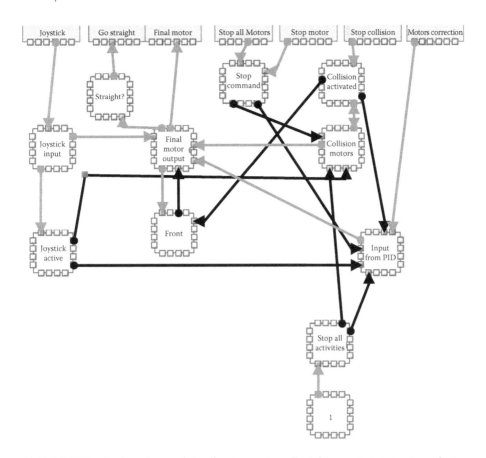

FIGURE 4.16 An iqr scheme of the director process. Each box represents a group of neurons. The group Final Motor Output outputs the corresponding command to the robot's motors. Light gray arrows indicate excitatory connections and dark gray arrows indicate inhibitory connections.

distinguish it from other odors. In this way we are able to define if one odor is of interest or not, and therefore we are able to develop a form of attraction or repulsion to an odor.

The classifier process is illustrated in Figure 4.18. All necessary information is received by the 10-neuron group Odor_id. In the first five cells, basic odors are mapped, while the last five cells are reserved for future use. When the classifier detects an odor, it may need some time to "smell" in order to classify it. Thus, it sends to the system a stop command for a few seconds. The time frame of the stop command varies according to the signal of the odor detected. The stop command is passed to the Stop Motors group that commands the motors through the director. If the desired smell is detected (in this schema the desired smell is ammonia), it will activate the odor detected cell from the cast-and-surge process and elicit an attractive behavior of the robot to that smell.

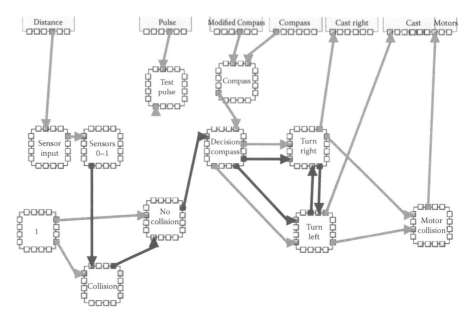

FIGURE 4.17 An iqr scheme of the collision detection process. Each box represents a group of neurons. The group Sensor Input receives the data acquired from the ultrasonic sensors, and the group Motor Collision outputs the corresponding command to the robot's motors. Light gray arrows indicate excitatory connections and dark gray arrows indicate inhibitory connections.

4.3.4.4.4 Cast and Surge

The cast-and-surge process is responsible for demonstrating the basic behavior of the robot, based on the complex behavior of the male moth when it is trying to locate the female. The male moth exhibits a specific upwind behavior called surge when it detects a pheromone plume and a crosswind behavior called cast when it loses track of the plume. We have managed to implement the same behavior on our robot by switching from casting to surging every time it encounters the desired odor.

As indicated in Figure 4.19, the robot switches between behaviors (casting, surging), depending on the activity displayed from the neuron odor detected. This neuron receives information from the process classifier, and it is activated if the target odor is detected. Thus, the default behavior of the robot is casting, and it is always active; when the target odor is detected and the odor detected displays activity, Begin Surge is enabled and Begin Cast is inhibited. In this way, we exclude the possibility of having activated at the same time both cast-and-surge actions.

4.3.5 Experimental Protocol

4.3.5.1 Locomotion and Maneuverability

To assess the maneuverability and locomotion of the robotic platform we placed the robot inside the testing arena with no odors present to perform a simple cast.

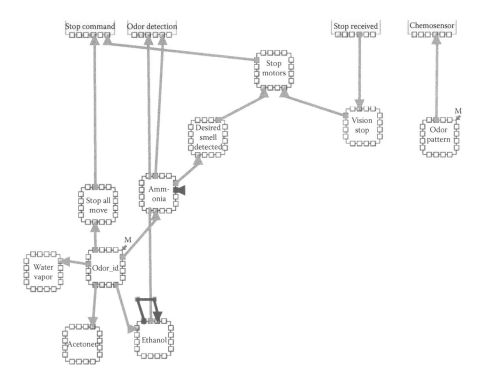

FIGURE 4.18 An iqr scheme of the classifier process. Each box represents a group of neurons. The group Odor_id receives the data acquired from the classifier. Light gray arrows indicate excitatory connections and dark gray arrows indicate inhibitory connections.

4.3.5.2 Static Classification

To appraise the classification performance, we divided the experimenting arena in a grid of points with 50 cm resolution. The robot was placed in each point facing upwind, for 1 min without moving, with only the classification model running. The experiment consisted of two odors where ethanol 20% and ammonia 5% were present. Measurements of the identified odors during that period of time were taken, and an odor map was reconstructed for the two different compounds separately. We called the resulting measurement classifications per minute (CPM), which represents the number of simulation cycles the classification model identified an odor during the 1 min period at each point.

4.3.5.3 Overall System

To evaluate the integration of both classification and localization models in our system the robot was placed inside the experimental arena in front of the ventilators facing upwind. The task was to perform a localization task in the presence of a target chemical compound, placed in the other end of the arena. Our aim was to see if the robot is able to correctly locate the source of the desired chemical compound when only the desired odor is present or when the desired odor has a chemical compound

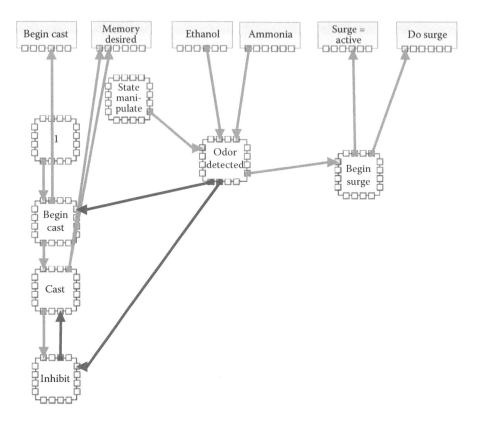

FIGURE 4.19 An iqr scheme of cast-and-surge process. Each box represents a group of neurons. The group Odor detected receives information from the classifier regarding the detection of the target odor. If it is detected, surge is activated and cast is inhibited. Light gray arrows indicate excitatory connections and dark gray arrows indicate inhibitory connections.

acting as a distracter (Figure 4.20). The two odors used were ethanol and ammonia in various concentrations, as shown in Tables 4.1 and 4.2.

4.4 RESULTS

4.4.1 Cast Performance—Sensor Validation

When the odor is not present, the default action is casting—a crosswind zigzag. In Figure 4.21 we can observe the robot's trajectory when no odor is present, and it is therefore casting. Our results show a correct crosswind casting movement as no chemicals are detected. To perform a complete cast in the whole arena, the robot needed 4 min and 51 s.

We also assessed the instrumentation capabilities of the robot. Two key sensors that we needed to assess were the compass and the wind vane. The compass outputs the direction of the robot in a unique cell, and therefore number, where each cell represents 10° of the robot's direction. Just like the static classification assessment, we recorded for each point the readings of the compass and compared them with

FIGURE 4.20 Image showing the robot while performing an odor localization and classification task inside the wind tunnel. Two odor sources are placed in the end of the wind tunnel. In this experiment, we used ethanol 20% (left) as a target odor and ammonia 5% (right) as a distracter.

TABLE 4.1
Concentrations Used for Assessing the System with One Odor and No Distracter

Single Odor, No Distracter	
Target: Ethanol	Target: Ammonia
20%	5%
10%	2%
5%	1%

Note: For each concentration, 10 trials were made.

TABLE 4.2
Concentrations Used When Assessing the System with Two Odors

Single Odor with a Distracter			
Target: Ethanol		Target: Ammonia	
Ethanol	Ammonia	Ethanol	Ammonia
20%	5%	20%	5%
10%	5%	20%	2%
5%	5%	20%	1%

Note: One odor acts as the target odor and the other as a distracter. In each case, the distracter remains stable and what varies is the concentration of the target odor.

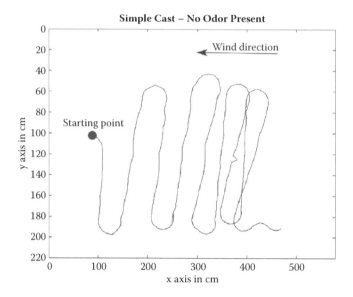

FIGURE 4.21 Image of the trajectory of the robot when no odor is present (casting). The starting point of the robot, marked by the circle, is located in the central point of the arena, in front of the ventilators facing upwind.

the simulated compass acquired from the AnTS tracking system. The heat map in Figure 4.22 displays the deviations from the readings of the compass compared to the readings that it should be displaying. The areas in red display highest deviation, and the areas in blue display the lowest deviation. As the compass is quite sensitive to magnetic fields, it is possible that the readings are affected by electric wires that are passing under the wind tunnel. Furthermore, we observe an excess deviation (of more than 150°) close to the ventilators, which can be explained by the ventilators themselves, as a magnetic field may be created by their movement. For this reason we have concluded that we will not be able to run successful experiments using the readings of the compass, and we had to simulate the compass from the AnTS tracking system.

Finally, we wanted to evaluate the readings of the wind vane. As we observed, when the robot is moving, the compass is affected by the robot's movement, and thus outputs false readings. We conducted experiments when the robot stood still in each point in the wind tunnel facing upwind. The wind vane outputs the direction of the wind in relation to the robot as a unique number, which is equivalent to a unique cell in the iqr system. We compared the wind vane's output with the simulated wind vane data acquired from the tracking system; Figure 4.23 displays a great deviation from the readings of the wind vane.

4.4.2 ODOR MAPS—CLASSIFICATION

The odor maps reconstructed from the static experiments can be seen in Figures 4.24 and 4.25. The X and Y axes represent the surface of the tunnel in centimeters, while

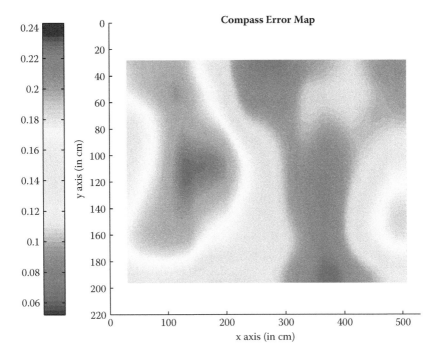

FIGURE 4.22 **(See color insert.)** Heat plot illustrating the deviation of the compass readings in comparison to the simulated compass. The areas of high deviation are marked in red, and of low deviation in blue.

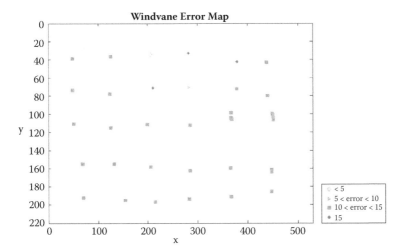

FIGURE 4.23 Error map of the wind vane. On each point marked the robot was facing upwind.

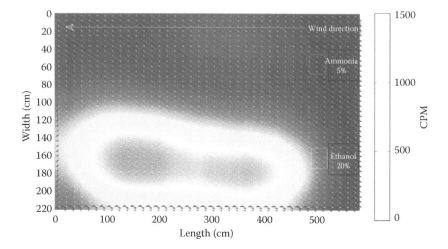

FIGURE 4.24 Odor map—ethanol detected: CPM in this case are considered as the number of cycles in which the classifier detected ethanol, with both odors in the air. The concentration used was 20% of ethanol in water.

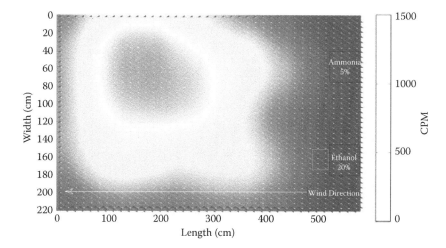

FIGURE 4.25 Odor map—ammonia detected: CPM in this case are considered as the number of cycles in which the classifier detected ammonia, being both odors in the air. The concentration used was 5% of ammonia in water.

the color scale indicates the classifications per minute value read in that position. We observe that the odor plume reconstructed corresponds to the right side of the tunnel where the source was placed.

At a first glance we can see that the detection of ammonia is much better than that of ethanol. While false readings for ammonia are always lower than 500 CPM, those for ethanol situate themselves in values around 1000 CPM. This translates into a lower success rate of localization of ammonia, since it is confused with ethanol as much as two times the other way around when working together with the localization model.

In any case the readings in CPM for the correct odor present were always higher than for the wrong one, converting the system in a positive classifier.

4.4.3 LOCALIZATION ASSESSMENT

We defined as successful the trial in which the robot reaches the target odor source while surging and the distance between the robot and the odor source does not exceed the length of the robot. A trial is successful even if the robot reaches the odor source laterally, as long as it is surging. We have come to this decision due to the fact that there is a small error in the compass of the robot (around $20°$), and there were cases where the robot was surging and due to either the location of the robot or the compass error, it reached the odor source laterally.

The robot had to perform the cast-and-surge task in two conditions: with the presence of only the target odor, and the presence of the target odor as well as a distracter. For the single odor set of experiments, we conducted in total 30 trials for ethanol and 30 trials for ammonia (10 trials for each concentration). Our results show that we had an overall success rate of 80% for ammonia and 86% for ethanol. Individually, the robot was able to correctly locate the source of ammonia with a percentage of 70% for 1%, 80% for 2%, and 90% for 5%. Similar results are found for ethanol, with a rate of 90% for ethanol 5%, 80% for 10%, and 90% for ethanol 20%. In Figure 4.26 we can see the trajectory of the robot during a successful trial for ethanol and ammonia, respectively.

The same task had to be performed in the presence of both a target odor and a distracter. In total we conducted 30 trials when the target was ethanol (10 trials for each concentration) and ammonia 5% was the distracter, and 30 trials when ammonia was the target and ethanol 20% was the distracter. Our results show an overall success rate of 90% for ammonia and 80% for ethanol. In the case of ammonia, there was a constant percentage of 90% of correct localizations for all three concentrations, whereas in the case of ethanol, there were 70, 80, and 90% successful localizations for 5, 10, and 20%, respectively. Figure 4.27 illustrates examples of correct localizations for ethanol and ammonia with the presence of a distracter.

Our results indicate a success rate that stands well above the random success rate of 60% for both cases of one single odor and one target odor with a distracter. These results suggest that our models are able to not only classify correctly a chemical compound but also locate its source.

4.5 CONCLUSIONS

In this chapter we have demonstrated the implementation of odor localization and classification models on an autonomous robot. The biological system on which we based our system is the male moth. Our main goal was to design a novel robotic system able to employ moth-like chemo-search strategies. Therefore, the localization model follows the same principles of the so-called cast-and-surge behavior of the male moth when it is trying to locate its mate. The classification model is based

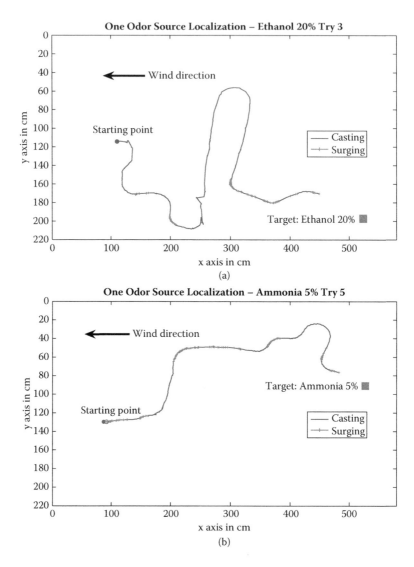

FIGURE 4.26 Image of a successful trial of the robot when it is locating ethanol 20% (a) and ammonia 5% (b). The starting point of the robot is marked by a circle and the location of the odor source is marked by a square. The moments where the robot has classified the desired odor and is therefore surging are marked with dashes.

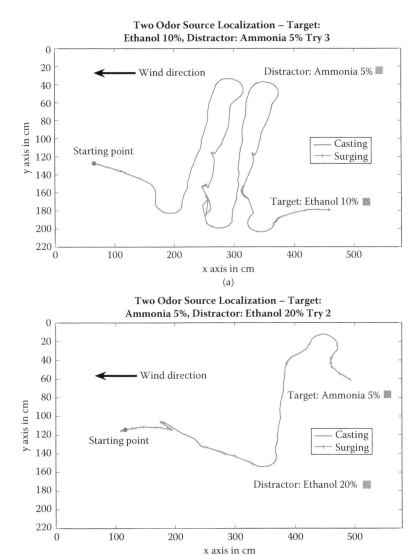

FIGURE 4.27   Image of a successful trial of the robot when it is locating ethanol 10% (a) and ammonia 5% (b) in the presence of a distractor. The starting point of the robot is marked by a circle and the location of the odor source is marked by a square. The distractor's position is labeled and marked by a square. The moments where the robot has classified the desired odor and is therefore surging is marked with dashes.

on the underlying neural structures of the first stages of the insect's olfactory path. Following the structure of the antenna, antennal lobe, and mushroom body, we have shown that the use of the technique of temporal population code (TPC) facilitates the discrimination of odors and results of actual use when processing real-time signals, and may also be present in the insect's brain. Both localization and classification models were implemented using the neuronal simulator iqr.

To maximize autonomy of the robotic system, we created a custom-made robotic platform with an embedded computer. Early experiments showed that the platform's initial design was not favoring maneuverability, control of the movements, and speed. We therefore redesigned the platform, introducing a reduction gear system to achieve optimal speed. From the first set of experiments, we observed an offset in the readings of both the custom-made wind vane and the compass. The wind vane was too sensitive when the robot was moving and not that accurate when the robot was still, while the compass may have been affected by the electricity wires that are passing under the arena of the wind tunnel. To overcome these problems, we simulated the robot's orientation and the airflow direction, based on information received by the tracking system.

We have shown that our system is able to perform a moth-like behavior. In the absence of an odor, the robot is casting, whereas when an odor is created, it switches to surge. In almost all trials, we have observed both types of behaviors, cast and surge, in the presence of a single target odor. Similar behavior is observed in the presence of both a target odor and a distracter, where the robot successfully identifies and locates the target odor, ignoring the distracter, suggesting that our model is quite similar to the techniques employed by the male moth.

Although there have been some early steps in implementing multimodal techniques to the existing system, by introducing vision and landmark navigation, further refinements and experiments of the model should be made. In this way, our system will not only be able to successfully identify and locate an odor and its source, but it will also be able to navigate through a complex background of visual landmarks, remember the path it followed by recognizing specific visual cues, and successfully return to its "nest." Thus, a robot's task may be to locate the source of a leakage of a hazardous gas and then return to its starting point safely.

ACKNOWLEDGMENTS

Supported by the European Community's Seventh Framework Program (FP7/007-2013) under grant agreement 216916. Biologically inspired computation for chemical sensing (NEUROCHEM).

REFERENCES

Baker, T. C., and Haynes, K. F. 1987. Manoeuvres used by flying male oriental fruit moths to relocate a sex pheromone plume in an experimentally shifted wind-field. *Physiological Entomology* 12: 263–279.

Beccherelli, R., Zampetti, E., Pantalei, S., Bernabei, M., and Persaud, K. C. 2009. Very large chemical sensor array for mimicking biological olfaction. *Olfaction and Electronic Nose: Proceedings of the 13th International Symposium on Olfaction and Electronic Nose* 1137(1): 155–158.

Bermudez i Badia, S., Bernardet, U., Guanella, A., Pyk, P., and Verschure, P. F. 2007. A biologically based chemo-sensing UAV for humanitarian demining. *International Journal of Advanced Robotic Systems* 4(2): 187–198.

Bermúdez i Badia, S., and Verschure, P. F. M. J. 2009. Learning from the moth: A comparative study of robot based odor source localization strategies. *AIP Conference Proceedings* 1137: 163–166. DOI: http://dx.doi.org/10.1063/1.3156498.

Bernardet, U., and Verschure, P. F. M. J. 2010. iqr: A tool for the construction of multi-level simulations of brain and behaviour. *Neuroinformatics* 8: 113–134.

Christensen, T. A., Waldrop, B. R., Harrow, I. D., and Hildebrand, J. G. 1993. Local interneurons and information processing in the olfactory glomeruli of the moth *Manduca sexta*. *Journal of Comparative Physiology A: Neuroethology, Sensory, Neural, and Behavioral Physiology* 173(4): 385–399. DOI: 10.1007/BF00193512.

Distante, C., Indiveri, G., and Reina, G. 2009. An application of mobile robotics for olfactory monitoring of hazardous industrial sites. *Industrial Robot: An International Journal* 36(1): 51–59.

Gomez-Marin, A., Duistermars, B., Frye, M. A., and Louis, M. 2010. Mechanisms of odor-tracking: Multiple sensors for enhanced perception and behavior. *Frontiers in Cellular Neuroscience* 4(6).

Grasso, F. W., Consi, T. R., Mountain, D. C., and Atema, J. 2000. Biomimetic robot lobster performs chemo-orientation in turbulence using a pair of spatially separated sensors: Progress and challenges. *Robotics and Autonomous Systems* 30(1–2): 115–131.

Hansson, B. S. 2002. A bug's smell research into insect olfaction. *Trends in Neurosciences* 270–224.

Knüssel, P. 2006. Dynamic neuronal representations of static sensory stimuli. PhD thesis, ETH, Switzerland.

Knüssel, P., Carlsson, M. A., Hansson, B. S., Pearce, T. C., and Verschure, P. F. M. J. 2007. Time and space are complementary encoding dimensions in the moth antennal lobe. *Network (Bristol, England)* 18(1): 35–62.

Kohnotoh, A., and Ishida, H. 2008. Active stereo olfactory sensing system for localization of gas/odor source. In *Proceedings of the 2008 Seventh International Conference on Machine Learning and Applications (ICMLA '08)*, 476–481. Washington, DC: IEEE Computer Society.

Kowaldo, G., and Russell, R. A. 2008. Robot odor localization: A taxonomy and survey. *International Journal of Robotics Research* 27(8): 869–894.

Kuwana, Y., Nagasawa, S., Shimoyama, I., and Kanzaki, R. 1999. Synthesis of the pheromone-oriented behaviour of silkworm moths by a mobile robot with moth antennae as pheromone sensors. *Biosensors and Bioelectronics* 14(2): 195–202.

Laurent, G. 1999. A systems perspective on early olfactory coding. *Science* 286(5440): 723–728.

Lilienthal, A., and Duckett, T. 2003. Experimental analysis of smelling Braitenberg vehicles. In *Proceedings of the IEEE International Conference on Advanced Robotics (ICAR)*, 20031, Coimbra, Portugal, 375–380.

Live Debian. Official website for Debian Live. http://live.debian.net/.

Lopez, L. L., Vouloutsi, V., Escuredo Chimeno A., Marcos, E., Bermúdez i Badia, S., Mathews, Z., Verschure, P. F. M. J., Ziyatdinov, A., and Perera i Lluna, A. 2011. Moth-like chemo-source localization and classification on an indoor autonomous robot. In *On biomimetics*, ed. L. D. Pramatarova. InTech. http://www.intechopen.com/books/on-biomimetics/moth-like-chemo-source-localization-and-classification-on-an-indoor-autonomous-robot.

Loutfi, A., and Coradeschi, S. 2008. Odor recognition for intelligent systems. *IEEE Intelligent Systems* 23(1): 41–48.

Mykytowycz, R. 1985. Olfaction—A link with the past. *Journal of Human Evolution* 14(1): 75–90.

Pearce, T. C., Chong, K., Verschure, P. F. M. J., i Badia, S. B., Carlsson, M. A., Chanie, E., and Hansson, B. S. 2004. Chemotactic search in complex environments. *Electronic Noses & Sensors for the Detection of Explosives,* 181–207. Vol. 159 of NATO Science Series II: Mathematics, Physics and Chemistry. Dordrecht, The Netherlands: Springer.

Purves, D., Augustine, G. J., Fitzpatrick, D. et al., eds. 2001. *Neuroscience.* 2nd ed. Sunderland. MA: Sinauer Associates.

Pyk, P., Bermudez i Badia, S., Bernardet, U., Knusel, P., Carlsson, M., Gu, J., Chanie, E., Hansson, B. S., Pearce, T. C., and Verschure, P. F. M. J. 2006. An artificial moth: Chemical source localization using a robot based neuronal model of moth optomotor anemototactic search. *Autonomous Robots* 20(3): 197–213.

R Development Core Team. 2011. *R: A language and environment for statistical computing.* Vienna, Austria: R Foundation for Statistical Computing.

R package chemosensors. https://r-forge.r-project.org/projects/chemosensors/.

Rachkov, M. Y., Marques, L., and de Almeida, A. 2005. Multisensor demining robot. *Autonomous Robots* 18(3): 275–291.

Schleef, D., Hess, F. M., and Bruyninckx, H. 2003. *The control and measurement device interface handbook.* http://www.comedi.org/doc/.

Todd, J. L., and Baker, T. C. 1999. Function of peripheral olfactory organs. In *Insect olfaction,* ed. B. S. Hansson, 67–96. Berlin: Springer-Verlag.

Trincavelli, M., Reggente, M., Coradeschi, S., Loutfi, A., Ishida, H., and Lilienthal, A. J. 2008. Towards environmental monitoring with mobile robots. *Intelligent Robots and Systems. IROS 2008.* IEEE/RSJ International Conference, 22–26, 2210–2215. Nice, France.

Waldrop, B., Christensen, T. A., and Hildebrand, J. G. 1987. GABA-mediated synaptic inhibition of projection neurons in the antennal lobes of the sphinx moth, *Manduca sexta. Journal of Comparative Physiology A: Sensory, Neural, and Behavioral Physiology* 161(1): 23–32.

5 Reactive and Cognitive Search Strategies for Olfactory Robots

Dominique Martinez and Eduardo Martin Moraud

CONTENTS

5.1 INTRODUCTION

Tracking scents and odor sources is a major challenge in robotics, with applications to the localization of chemical leaks, drugs, and explosives (Russell 1999). Nowadays, animals are commonly used in safety and security tasks because of their excellent smell detection capabilities. Examples include dogs and honeybees (Rains et al. 2008). However, using animals to sniff specific odors related to bombs or explosives has several drawbacks. On top of the hazards of such endeavors, animals like dogs show behavioral variations and changing moods. They get tired after extensive work and require frequent retraining as their performance decreases over time. As an alternative, could we envision using olfactory robots to advantageously replace animals for these tasks?

At short distances from the source, olfactory search methods inspired by bacterial chemotaxis provide acceptable solutions for navigating a robot. Bacteria like *Escherichia coli* direct toward chemo-attractants by climbing a concentration gradient. They alternate between periods of straight swims called runs and random

153

FIGURE 5.1 Different search strategies can be used depending on the type of olfactory cues available, concentration gradient (a) or intermittent odor patches (b and c). (a) Climbing a concentration gradient is possible at short distances from the source by modulating the trajectory curvature according to the difference in concentration between left and right sensors (Hugues et al. 2003; Martinez et al. 2006). (b) Reactive search strategies take inspiration from the odor-modulated anemotaxis of male moths (as reviewed extensively in Murlis et al. 1992; Kaissling 1997; Vickers 2006). A spiral-surge strategy combines upwind surge in the presence of the odor with spiraling in its absence (Hayes et al. 2002; Lochmatter et al. 2008). (c) Cognitive search strategies like infotaxis (Vergassola et al. 2007) are based on probabilistic inference using an internal model of the world. The searcher updates its belief (probability map of the source distribution) according to past actions and sensory observations (history of detection and nondetection events).

reorientations called tumbles (Berg 2003). Such biased random walks have been implemented on real robots (Lytridis et al. 2006; Russell et al. 2003; Marques et al. 2006), yet with a limited success, mainly because of the use of a single odor sensor. Unlike bacteria, other animals may use instantaneous gradient information assessed by comparing the responses of spatially separated chemosensors. This claim is supported by experiments showing that unilateral lesions, like the blockage of the nasal airflow in one nostril in rats or the ablation of one of the antennae in crayfish, impair odor source localization (Kraus-Epley and Moore 2002; Rajan et al. 2006; McMahon et al. 2005). Autonomous olfactory robots using bilateral comparison include the robotic "lobster" for tracking saline plumes in water (Consi et al. 1995; Grasso et al. 1997; Grasso 2001), the Braitenberg olfactory robot (Lilienthal and Duckett 2003), and our robot (Hugues et al. 2003; Martinez et al. 2006), whose trajectory curvature was constantly modulated by the difference in concentration between left and right sensors (Figure 5.1a). A prerequisite to all the aforementioned chemotactic robots is the existence of a relatively smooth concentration gradient. The experiments we performed revealed that a concentration gradient can effectively be measured (see Figure 5.4, left in Martinez et al. 2006), but only when the robot moves slowly (2.5 cm/s) and near to the source (search area limited to 3 m^2). As chemotactic search strategies are applicable only in the vicinity of the source, we considered the possibility of exploring the environment by using vision, in addition to olfaction (Martinez and Perrinet 2002). An important limitation nevertheless is that odor source candidates need to be identifiable from visual features.

Far from the source, the concentration landscape of an odor, called a plume, is very heterogeneous and unsteady, and consists of sporadically located patches (Weissburg 2000; Roberts and Webster 2002). Even at moderate distances (order 10 m), detections become sporadic and only provide cues intermittently. Given this discontinuous flow of information, how then can we efficiently navigate a robot toward the source over moderate or large distances (order 100 m)? It is well known that insects such as male moths successfully locate their mates over distances of hundreds of meters. To do so, male moths adopt a typical behavior (for reviews see, e.g., Murlis et al. 1992; Kaissling 1997; Vickers 2006). Upon sensing a pheromone patch, they surge upwind, and when the odor information vanishes, they perform an extended crosswind casting until the plume is reacquired. This strategy has the advantage of being purely reactive; i.e., actions are completely determined by current perceptions, that is, surge upwind upon sensing a pheromone patch and cast crosswind when odor information vanishes. Such reactive methods have been simulated or implemented on robots in various forms (Kuwana et al. 1999; Pyk et al. 2006; Balkovsky and Shraiman 2002). An efficient variant is the spiral-surge strategy (Hayes et al. 2002; Lochmatter et al. 2008) that combines upwind surge in the presence of the odor with spiraling in its absence (Figure 5.1b). Yet performance of reactive strategies at distances beyond 100 m, when the reacquisition of the plume becomes very unlikely, is unclear. Reactive casting-surge methods address the search problem only from an imitation perspective; i.e., they mimic the choices performed by animals through a rule-based approach, regardless of the underlying mechanisms from which the behavior emerges. This biomimetic approach raises the question of how well reactive strategies may be adapted to new environmental conditions as those occurring when the distance from the source increases.

For those conditions, a more sophisticated method, infotaxis, was proposed recently (Vergassola et al. 2007; Martinez 2007). Infotaxis relies on Bayesian inference to maximize information gain, and exploits the expected distribution of odor encounters. It involves a period of exploration during which the searcher builds a probabilistic map of the source location. As the searcher accumulates information, the map becomes sharper and its entropy—which reflects the uncertainty about the location of the source—decreases. Because the expected search time is determined by the uncertainty of the belief, the searcher moves so as to maximize the expected reduction in entropy, and therefore the rate of information acquisition. Maximizing information gain entails a competition between two actions, exploitation and exploration. The former drives the searcher toward locations where the probability of finding the source is high. The latter favors motion to regions with lower probabilities of source discovery but high rewards in terms of information gain. Infotaxis is a cognitive strategy in the sense that an internal model of the world is built from past detections and actions so that memory and learning play a crucial role (Figure 5.1c).

Advantages, disadvantages, limitations, and biological plausibility of cognitive and reactive search strategies are largely unknown and remain to be quantified. In this chapter, we review the two approaches and report comparisons based on simulations and robotic experiments. To our knowledge, only two studies have considered a robotic implementation of infotaxis: Lochmatter (2010) and Moraud and Martinez (2010).

5.2   REACTIVE AND COGNITIVE SEARCH STRATEGIES: SPIRAL-SURGE AND INFOTAXIS

The reactive and cognitive search strategies described in this section are based on a single odor sensor providing binary information about the presence or absence of the odor, and thereby ignore the instantaneous level of concentration. The uselessness of concentration in tracking turbulent plumes is justified by several studies revealing that concentration can still be high within filaments far away from the source (Jones 1983), and that the local concentration gradient does not always point toward the source (Murlis et al. 2000).

5.2.1 REACTIVE SEARCH: SPIRAL-SURGE

Most reactive search strategies take inspiration from the flight of male moths localizing a conspecific female from far away. This is one of the most remarkable and best-studied olfactory searching behaviors known in nature (for a review see, e.g., Murlis et al. 1992; Kaissling 1997; Vickers 2006). Male moths flying within a pheromone plume exhibit a characteristic odor-modulated anemotactic response. They surge upwind upon sensing a pheromone patch and perform an extended crosswind casting when odor information vanishes. The difficulty of translating these ideas into an algorithm is to be able to specify the casting phase (the surge phase being simply a straight movement in the upwind direction). Balkovski and Shraiman (2002) proposed a casting algorithm in which the searcher zigzags upwind and the parameters of the turns, amplitude and upwind drift, are adapted to statistical properties of the flow. Because the searcher moves solely upwind, however, the source can be missed, and there is no chance to localize it downwind. In practice, male moths do not always fly upwind. For example, foraging or appetitive flights of male moths, prior to initial contact with the odorant, have been shown to be either downwind (Reynolds et al. 2007) or with no preferred orientation with respect to the wind flow (Cardé et al. 2012). During casting, the silkworm moth *Bombyx mori* first performs zigzag turns across the wind in the hope of regaining contact with the odor, but then switches to circular motion patterns (Figure 5.2a). Inspired by the looping behavior of the silkworm moth, the spiral-surge algorithm (Hayes et al. 2002) performs an Archimedean spiral in the casting phase (Figure 5.2b) and yields significantly better results in practice than a zigzagging casting algorithm (Lochmatter et al. 2008).

5.2.2 COGNITIVE SEARCH: INFOTAXIS

In reactive search strategies, behavioral patterns like surge and casting are preprogrammed, and the actions of the agent are completely determined by the current perceptions. The cognitive search strategy infotaxis (Vergassola et al. 2007), on the contrary, is based on reinforcement learning and fully exploits the capabilities of autonomous on-line learning. Driven by a decision-making strategy that efficiently combines exploration of the surrounding and information exploitation, the agent discovers the direction leading to the source by interacting iteratively with the environment. Infotaxis is built around two core components: uncertainty modeling and decision making. The former is achieved through an internal description of the world (physical description of turbulent transport), which is used by the latter to interpret odor encounters and infer the likelihood of the source to be at a given location. We outline hereafter these components.

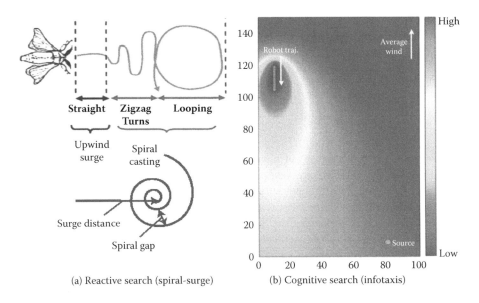

(a) Reactive search (spiral-surge)   (b) Cognitive search (infotaxis)

FIGURE 5.2 **(See color insert.)** Reactive and cognitive search strategies. (a) Spiral-surge (reactive search) is inspired by the odor-modulated anemotactic response of male moths in pheromone plumes. Top: A typical trajectory of a silkworm moth consists of two phases, an upwind surge in the presence of the odor and a casting phase in its absence. During casting, the moth performs first zigzag turns across the wind in the hope of regaining contact with the odor, and then switches to circular motion patterns. (Reprinted with modification from Kanzaki, R., Ando, A., Sakurai, T., and Kazawa, T., *Adv. Robotics*, 22(15), 1605–1628, 2008. With permission.) Bottom: In spiral-surge, the casting phase is modeled as an Archimedean spiral with a gap parameter (Hayes et al. 2002; Lochmatter et al. 2008). The surge phase requires knowing the wind direction and has a single parameter, the surge distance. (b) Infotaxis (cognitive search). Example of belief (probability map) for the location of the source after 10 steps (red dots). No cues are detected in that time. Locations in front of the agent become less probable as the agent navigates forward without encounters, thereby increasing the likelihood of locations on the sides. The shape of the belief (Gaussian-like) is inferred from the physical description of how cues spread in the environment when transported by the wind.

5.2.2.1 Uncertainty Modeling: Internal Representation of Turbulent Transport and Probabilistic Belief

The agent employs an internal description of how cues spread in the environment. This representation provides information of what to expect given what is perceived, and is essential to interpret interactions with the surrounding. In the case considered in Vergassola et al. (2007) the internal model corresponds to the spatiotemporal profile of odor plumes as derived from the equations of diffusion-advection. However, since odor dispersal in open environments is subject to high degrees of turbulence and randomness, a detailed description of the environmental dynamics is difficult to achieve. Instead, a statistical model is employed, in which the odor cues perceived at location r, during a discrete-time interval δt, are considered to be independently sampled from a Poisson distribution with parameter $\lambda(r, s_0) = R(r, s_0)\delta t$, where $R(r, s_0)$ is the time-averaged detection rate given a source located at position s_0. Hence,

$$p(k = 0 \ detection \mid s_0) = \exp\left[-R(r,s_0)\delta t\right]$$

$$p(k = n \ detections \mid s_0) = \left[R(r,s_0)\delta t\right]^n \Big/ n! \ \exp\left[-R(r,s_0)\delta t\right]$$

Based on the previous model of odor encounters, a probabilistic belief is built (Figure 5.2, right) given the trace of past detections. This is done in a similar way to Pang and Farrell (2006). The belief is a grid-based map of the environment (closed-world assumption is adopted) that is constructed by inferring the probability of the source to be located in s_0 given the trace of past detections $\wp = \{(r_0, t_0), (r_1, t_1), \ldots, (r_n, t_n)\}$. Because the agent did not detect at any moment other than $t_0 \ldots t_n$, nondetections also provide information and are used to derive the belief. Note also that the path followed between the initial and the current state is implicitly encoded in $\wp$. Updating the probability map corresponds to deriving the posterior probability of the random variable s_0 from the n observations in $\wp$, and is thus directly obtained through Bayesian inference. Since cues are expected to follow a Poisson distribution and are assumed to be independent and identically distributed, the belief up to time t is calculated as

$$p_t(s_0) \propto \left(\prod_{\forall r_i \in \wp} R(r_i, s_0) \ \delta t\right) \exp\left(-\sum_{j=0}^{t} R(r_j, s_0)\delta t\right)$$

where r_i represents the locations of the n detections observed in $\wp$ at times t_i, $i = 1, \ldots, n$. Such a function is iteratively updated as the agent moves.

5.2.2.2 Decision Making

The approach employed by infotaxis to decide where to move next is the key point of its robustness. Unlike classical navigation methods (maximum likelihood or maximum a posteriori), infotaxis does not target the most likely location for the source. Instead, the agent chooses to move to the new location $\hat{r}$, which maximally reduces uncertainty:

$$\hat{r} = \arg\max_{r_j} \left(S_{t+1} - S_t\right)$$

where S_t is the Shannon entropy at the current step t: $S_t = E[-\ln p_t(s_0)]$, with E the expectation taken over the source location s_0. This decision making naturally conveys a balance between how much is currently known and how much is yet to be discovered, or in other words, between *exploration* and *exploitation*. The variation of entropy consists of two parts:

$$S_{t+1} - S_t = p_{t+1}(r_j)(0 - S) + \left[1 - p_{t+1}(r_j)\right]E(\Delta S_j)$$

The first term evaluates the probability $p_{t+1}(r_j)$ of finding the source in the next step when moving from r to r_j—in which case the entropy becomes zero. The second term computes the amount of knowledge $E(\Delta S_j)$ gathered when the source is not found with probability $1 - p_{t+1}(r_j)$.

The first term corresponds to the exploitative choice, driving the searcher toward locations that maximize its (expected) chances of finding the source—regardless of other considerations. The second term represents the explorative decision, which favors motion to regions where the agent might detect new cues—regardless of whether the source is actually believed to be in that direction or not. This balance is essential for the strategy to be effective and provides the model with a robustness that makes it especially suitable for turbulent environments.

5.3 NUMERICAL COMPARISONS BETWEEN SPIRAL-SURGE AND INFOTAXIS

Using numerical simulations, we assess here the performance of spiral-surge and infotaxis in terms of complexity (simulation time), effectiveness (search time), and robustness (with respect to changes in environmental conditions). Both algorithms were implemented in Python (see appendix) and evaluated on the same computer (MacBook Pro, 2.2 GHz).

The parameters of infotaxis were identical to those of the statistical model of the odor plume used in simulations, with details given in Moraud and Martinez (2010). The environmental conditions were defined as follows: search area of 5×5 m^2 discretized as a 100×100 grid, source diffusivity $D = 1$ a.u., emission rate $R = 1$ a.u., lifetime of particles $\tau = 300$ a.u., wind speed $V = -2$ m/s, and sensor size $a = 1$ cm. The parameters of spiral-surge have been estimated to provide acceptable performance. In the simulations, we used a spiral gap of 30 cm and a surge distance of 50 cm. Examples of trajectories are depicted in Figure 5.3a and b for spiral-surge and infotaxis, respectively. In order to obtain statistically comparable results, we performed 150 runs of each algorithm with the source located at (50, 80) and the robot starting at (20, 20). The searcher was assumed to have reached its goal at one step from the source. In terms of effectiveness, infotaxis outperformed spiral-surge (search time = 724 ± 341 steps for infotaxis versus 1124 ± 703 steps for spiral-surge). In terms of complexity, spiral-surge was much less demanding (simulation time = 60 ± 29 s for infotaxis versus 0.2 ± 0.1 s for spiral-surge).

To assess robustness, we tested the two methods under varying environmental conditions (lifetime of particles τ in range 100–600 a.u. and wind speed V in range 0.5–3.5 m/s). Infotaxis was clearly more robust than spiral-surge, for which changes in environmental conditions induced more variations in the search time (Figure 5.3c and d). Infotaxis, however, may be highly dependent on a particular statistical model of the odor plume, as the likelihood for the source to be at a given location is inferred through an internal description of the environment, i.e., a physical description of how cues spread when transported away from the source.

Simulations performed so far were done with a continuous source. Females of some moth species are known to rhythmically extrude their pheromone glands

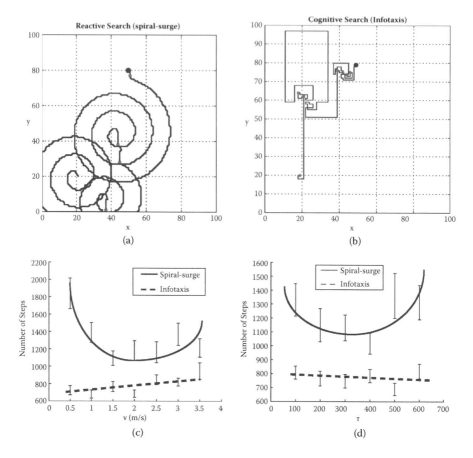

FIGURE 5.3 Performance comparison between reactive and cognitive plume tracking strategies. (a, b) Examples of simulated trajectories with correct modeling by the searcher (spiral-surge in a, infotaxis in b). The environment (5×5 m) is a grid-based model of 100×100 points. The source is located at (50, 80). The robot starting point is (20, 20). Wind blows downwards. The robot is assumed to have reached its goal at one step from the source (c, d). Performance of spiral-surge and infotaxis with incorrect modeling by the searcher. (c) The searcher considers a wind speed v different from the actual one ($v = 2$ m/s). (d) The searcher considers a lifetime of a particle different from the actual one ($\tau = 300$ a.u.).

(Baker et al. 1985). To assess the capacity of infotaxis to cope with real conditions such as those faced in biology, we considered simulations with a pulsed source model that rhythmically releases odor patches in the environment. Note that there was a mismatch between the pulsed source generator (pulse duration = 0.2 s, air gap between pulses = 1.3 or 4.8 s) and the continuous source model ($D = 1$ a.u., $\tau = 1.5$ a.u., $R = 2$ a.u., $V = -2.5$m/s, and $a = 1$cm) used to update the internal beliefs in infotaxis. We performed 60 simulation runs, 30 with fast pulses (air gap between pulses = 1.3 s) and 30 with slow pulses (air gap = 4.8 s). Percentages of downwind movements were low and not significantly different in both conditions (10.1 ± 4.2% for fast pulses versus 10.6 ± 7.5%). We found, however, that the searcher moves mainly upwind in the fast pulsed condition (52 ± 8% of upwind movements for fast pulses versus 43 ± 6% for slow pulses) and crosswind in the slow pulsed condition (38 ± 6% of crosswind movements for fast pulses versus 46 ± 8% for slow pulses). Typical infotactic trajectories under fast and slow pulsed conditions are shown in Figure 5.4. With fast pulses (Figure 5.4a), intervals of no detection between pulses are short enough to keep the searcher on exploitation. Each update sharpens the posterior distribution. The high-probability bump emerging in the wind direction induces the agent to move upwind. With slow pulses (Figure 5.4b), unexpected long periods of time with no odor encounter broaden the posterior distribution, hence compelling the agent to counterturn and explore the environment in large spirals.

5.4   ROBOTIC IMPLEMENTATION OF INFOTAXIS

We addressed the problem of verifying that infotaxis would be equally efficient under real experimental conditions as it was in simulation. One of the motivations is that computational models usually require simplifications or assumptions helping to make problems tractable. On the contrary, robotic agents are confronted to the real environment, and hence provide a test bed to assert complete and rigorous results. More importantly, robotic implementations also represent an essential step to ensure that algorithmic concepts can be implemented with the available technology and employed beyond computer simulations.

The key point at the core of infotaxis is the randomness of odor encounters. This randomness motivates the use of a cognitive approach that employs uncertainty minimization navigation techniques. Yet in simulations, cues were modeled through stochastic mathematical descriptions that assume independent and uncorrelated hits. In reality, however, an odor patch covers a certain volume and presents extended spatiotemporal characteristics. Even though inherently random, this structure will give rise to consecutive nonindependent cues (at the sensor sampling time). For infotaxis to be fully efficient, consecutive detections should not be overcounted. In Moraud and Martinez (2010), the posterior probability distribution was derived from a modified model that accounts for correlated hits, and built on transitions (from no detection to detection) rather than on single hits. In our implementation, this was achieved by means of an adaptive filter, calculated over a moving time window (Figure 5.5).

Furthermore, electronic sensors must be chosen so that the requirements of the model (in terms of sensitivity and speed) are met. Because the odor sensors we tested so far required long degassing times and saturated easily, they were unable to respond to the requirements of infotaxis. As an alternative, we chose to use heat sensors, which do not saturate easily and react at high speed. We note that the spatiotemporal distribution of heat is identical to that of odor, and thus no loss of accuracy is brought in by this adaptation.

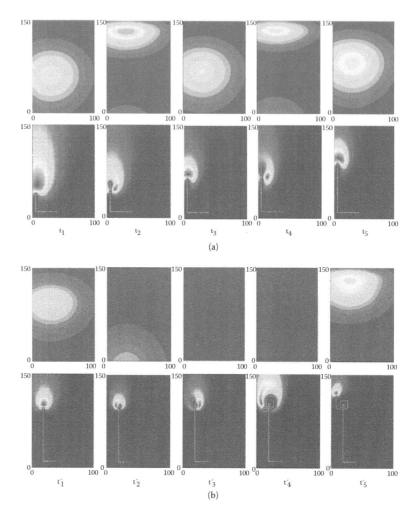

FIGURE 5.4 **(See color insert.)** Navigation patterns observed when infotaxis is confronted to a pulsed source (fast pulses in a, slow pulses in b). Top rows represent snapshots of the simulated environment at different times (pulsed source located in $r_0 = (25, 2)$, wind blows downward), and bottom rows are for the corresponding source distribution (belief function). False blue and red colors correspond to low and high probabilities, respectively. The path of the robot is superimposed to the map as consecutive red dots when there is no detection, and green dots for detections. (a) High-frequency pulsed patches provoke new detections before the robot starts spiraling. High probabilities are frequently updated and assigned to upwind locations (at times t_1, t_3, and t_5), hence pushing the agent forward. (b) Between pulses, long intervals of clean air—during which no detections arise—compel the agent to explore regions where previous detections were recorded. Probabilities updates take the form of concentric ellipses that spread as the robot navigates around them, as clearly seen at times t_2', t_3', and t_4'. (Note that such behavior—although with much smaller radius—may also be recorded for the fast-pulsed case when the agent is close to the source, due to an excessive amount of detections that push him to switch to exploitation mode.) (Reprinted from Moraud, E. M., and Martinez, D., *Front. Neurorobotics*, 4: 1, 2010, DOI: 10.3389/fnbot.2010.00001.)

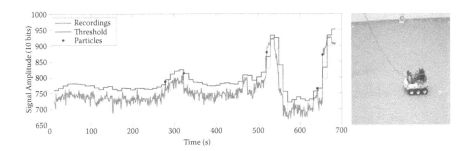

FIGURE 5.5 Real recordings (dark gray) at 10 Hz sampling frequency, and derived cues (black dots) to be employed by infotaxis when guiding the search of the robot represented at the right. A moving window (light gray) is used to filter the signal while preventing consecutive correlated hits from being overcounted, thereby ensuring that cues are appropriately derived from the sensor measurements. (From Moraud, E. M., and Martinez, D., *Front. Neurorobotics* 4: 1, 2010, DOI: 10.3389/fnbot.2010.00001.)

When compared to simulated results, identical distributions were obtained for the robot trajectories, both for the search time required until finding the source and for the number of encounters required (Figure 5.6), thus ensuring that its main properties are preserved when applied to reality. Note also that the internal model relied upon by the agent employs information of diffusion parameters (e.g., wind speed and direction, which in reality may vary over time and differ from the estimated ones). In the robotic experiments, the robustness of infotaxis was thus evaluated even with respect to inaccurate modeling, as the parameters were not fine-tuned or adapted on-line. Yet despite this discrepancy, the robot was able to find the source within reasonable time limits.

The biomimetic characteristics of the navigation were also preserved in our robotic implementation. Robot trajectories were shown to exhibit animal-like patterns such as extended crosswind or zigzag upwind. The track angle histogram also maintains a distribution similar to that observed in moths; details are given in Moraud and Martinez (2010).

5.5 PERSPECTIVES AND CONCLUSION

In this chapter we reviewed spiral-surge and infotaxis, two strategies for searching a source of information (e.g., chemical or heat) based on intermittent cues. Although both strategies may be understood in terms of exploration and exploitation (Balkovsky and Shraiman 2002; Vergassola et al. 2007), they have different grounds: probabilistic inference for infotaxis and bioinspiration for spiral-surge.

Infotaxis is a cognitive strategy in the sense that the next action is inferred through the use of an internal model built from past observations. On the opposite, spiral-surge is a reactive strategy in the sense that the next action (surge or spiral) is just computed from the current context (presence or absence of the stimulus). The parameters of the actions, surge distance and spiral gap, have to be tuned to the current environmental conditions. The robustness of both strategies was evaluated with

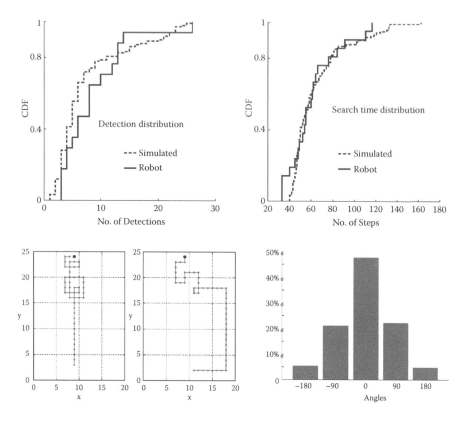

FIGURE 5.6 Comparison of robotic and simulated results: cumulative distribution of the number of steps until finding the source (right) and the number of cues required to reach the goal (left).

respect to inaccurate modeling by the agent. The parameters employed internally to guide the search were not fine-tuned or adapted over time, and could differ from the instantaneous characteristics of the surrounding. Despite this discrepancy, the agent guided by either infotaxis or spiral-surge reached the source within a reasonable time limit. Based on numerical simulations, we reported that infotaxis is more effective and more robust than spiral-surge.

The superior performance of infotaxis comes at the cost of increased computational requirements. Does this increased complexity compromise the use of infotaxis in real robots? The robotic implementation presented in this chapter demonstrated that this is not the case for the chosen grid-based map (100 × 100 with 5 cm/sector). Nevertheless, it may be necessary to make simplifications when the resolution of the grid (or the size of the covered surface) increases, as heavy computations would exponentially become a non-negligible burden incompatible with an on-board implementation. A future challenge would therefore be to simplify the infotaxis strategy of search in a suitable form for on-board, real-time implementation. Methods that replace the grid map for a continuous-space version of the algorithm (e.g., Barbieri et al. 2011) may prove more easily scalable. Likewise, methods that employ approximate inference for deriving the posterior beliefs may also provide faster solutions.

Again from a technical perspective, the presented strategies rely on ideal sensing capabilities. Yet, we tested several gas sensors based on tin oxide, and none of them suited our needs in terms of response time and sensitivity. They saturated at medium concentrations and require a long-lasting phase of degassing before they can react again. The difficulty to resolve individual odor patches with tin oxide gas sensors motivated us to test infotaxis on a transport model of heat identical to the one of smell. However, alternative solutions that exploit current advances in gas sensing technology (e.g., a miniature photoionization detector) may also give answer to the aforementioned requirements.

In reactive search strategies like spiral-surge, the actions of the agent are imposed by explicit rules of movements, e.g., upwind surge in the presence of the odor and casting in its absence. Behavioral patterns of male moths attracted by a sex pheromone seem rather stereotyped and might then be simply innate or preprogrammed. It has been suggested that surge and casting behaviors are instructed by particular types of descending neurons in the protocerebrum of the moth (Olberg 1983; Kanzaki et al. 1991, 1994). However, it might be too simplistic to consider the moth behavior as a hard-wired program. After all, a cognitive strategy like infotaxis also produces behavioral patterns resembling spiral-surge, as noticed, for example, in Figure 5.4b. Infotactic spiraling trajectories in the absence of the odor were also reported in other studies (Masson et al. 2009; Barbieri et al. 2011). The main difference with a reactive spiral-surge strategy is that these behaviors emerge naturally from the trade-off between exploration and exploitation at the core of the underlying cognitive spatial model (belief function). Yet, whether or not insects rely on cognitive maps for their navigation is an open question (Wehner and Menzel 1990; Collett and Collett 2002). Future work is thus needed to investigate under which form a cognitive search strategy like infotaxis could be implemented in a moth brain.

Interestingly, the search strategies presented in this chapter are general to searching a source of information based on intermittent cues. Indeed, the concept only requires certain features to work (i.e., cues encountered along the way, which provide information about the source given with a model of the environmental dynamics). Such methods may thus be extended to a variety of scenarios dealing with autonomous navigation and goal localization under uncertainty. A key point is that the goal is treated as a source of information that spreads cues in the environment in a way that can be modeled, and relied upon when reasoning. Examples may include rovers searching for footprints of biological activity or chemical compounds spread in the soil, as, for instance, in the framework of space exploration where efficiency and autonomy are essential for the success of the missions (Moraud and Chicca 2011).

5.6 APPENDIX: SOURCE CODE FOR INFOTAXIS AND SPIRAL-SURGE

5.6.1 INFOTAXIS

Infotaxis was implemented in two major classes, one that describes the environment and another that updates the likelihood map and computes the variation in entropy when moving to neighbor locations. The code is outlined below:

```python
import numpy
import donnees
from scipy import weave, factorial, cumsum
import matplotlib.pyplot as plt
import sys
class probMap:
"""A grid class that stores the details and solution of the
computational grid."""
    def __init__(self, V_y):
            self.dt = donnees.dt
            self.nx = donnees.quad_x
            self.ny = donnees.quad_y
            self.sautx = donnees.saut_x
            self.sauty = donnees.saut_y
            self.Vy = V_y
            self.lamda = numpy.sqrt((donnees.info_D*1.*donnees.
info_tau)/(1+ (self.Vy*self.Vy*donnees.info_tau*1.)/
(4.*donnees.info_D)))
            self.XaxisMatrix = numpy.zeros((self.ny,self.
nx),dtype = int)
            self.YaxisMatrix = numpy.zeros((self.ny,self.
nx),dtype = int)
            for i in range(self.ny):
                    self.XaxisMatrix[i,:] = numpy.
arange(0,self.nx)
            for i in range(self.nx):
                    self.YaxisMatrix[:,i] = numpy.
arange(0,self.ny)
            self.u = numpy.ones((self.ny, self.nx), dtype =
float)
            self.Den = 0
            self.S = 0
            self.old_u = self.u.copy()
            self.index = 0
    def detRate_grid(self, Xagent, Yagent):
            "Detection-rate grid for each point in the grid
given a position for the agent"
            # Matrix of size [ny,nx] with all the distances
between the agent an any other point of the grid
            module_dist = numpy.sqrt(numpy.power((Xagent-self.
XaxisMatrix)*self.sautx, 2) + \
```

```
                        numpy.power((Yagent-self.
YaxisMatrix)*self.sauty, 2))
            # Average concentrations given previous matrix
            conc = donnees.info_R*1./(4.*numpy.pi*donnees.
info_D*module_dist) * \
                        numpy.exp(-1.*(Yagent-self.
YaxisMatrix)*self.sauty*self.Vy/(2.*donnees.info_D)) *numpy.
exp(-1.*module_dist/self.lamda)
            conc[Yagent,Xagent] = 100
            # Detection-rate matrix
            detRate = 4.*numpy.pi*donnees.info_D*donnees.
info_a*conc
            return detRate
    def forward(self, Xagent, Yagent, Nparticules):
            "Calculate likelihood for each next location"
            detRate = self.detRate_grid(Xagent, Yagent)
            #Likelihood map
            u_new = numpy.exp(-detRate*self.dt) * numpy.
power(detRate*self.dt, Nparticules)
            # Update
            u_tot = self.u * u_new
            Den = numpy.sum(u_tot)
            return u_tot, Den
    def update(self, Xagent, Yagent,Nparticules):
            "Update values for self.u, self.Den, self.S"
            self.old_u = self.u.copy()
            self.u, self.Den = self.forward(Xagent, Yagent,
Nparticules)
            self.S = self.entropy(self.u, self.Den)
            self.index = self.index+1
    def entropy(self,u,Den):
            "Calculates the entropy"
            S = - 1.0 * numpy.sum(u/Den*(numpy.log(u) - numpy.
log(Den)), dtype = float)
            return S
    def plot(self):
            proba = self.u #/self.Den Avoid overflows... Plot
does not change anyway
            fig = plt.figure()
            ax = fig.add_subplot(111)
            cax = plt.pcolormesh(proba)
            plt.axis([0,self.nx,0,self.ny])
            ax.set_title('Probability grid map')
            cbar = fig.colorbar(cax,ticks = [numpy.min(proba),
numpy.max(proba)])
            cbar.ax.set_yticklabels(['Low', 'High'])
            fig.savefig('infotaxis'+str(self.index))
            plt.show()
    def poisson (self, Npoints, landa):
            increasingNbs = numpy.linspace(0, Npoints-1,num =
Npoints)
```

```
            poissonVals = numpy.power(landa,increasingNbs) *
numpy.exp(-landa) *1./factorial(increasingNbs)
            return poissonVals
    def varEntropy(self, Xneighbour, Yneighbour):
            "Variation in entropy when going to neighbor j"
            # Detection-rate grid if I go to the next neighbour
            detRate = self.detRate_grid(Xneighbour, Yneighbour)
            #Current prob grid
            prob = self.u/self.Den
            # Expected detection.rate at the neighbour
            h = self.dt * numpy.sum(prob * detRate)
            N = 5
            ro = self.poisson(N,h)
            # Forward variation of prob grid and entropy
            varS = 0
            for k in range(N):
                    u,den = self.forward(Xneighbour,Yneighbou
r,k)
                    varS = varS + ro[k]*(self.entropy(u,den)
- self.S)
            return varS
    def sizeof(self):
            size = sys.getsizeof(self.dt) \
                + sys.getsizeof(self.nx) \
                + sys.getsizeof(self.ny) \
                + sys.getsizeof(self.sautx) \
                + sys.getsizeof(self.sauty) \
                + sys.getsizeof(self.Vy) \
                + sys.getsizeof(self.lamda) \
                + sys.getsizeof(self.XaxisMatrix) \
                + sys.getsizeof(self.YaxisMatrix) \
                + sys.getsizeof(self.u) \
                + sys.getsizeof(self.Den) \
                + sys.getsizeof(self.S) \
                + sys.getsizeof(self.old_u) \
                + sys.getsizeof(self.index)
            for o in self.XaxisMatrix:
                    size = size + sys.getsizeof(o)
                    for o2 in o:
                            size = size + sys.getsizeof(o2)
            for o in self.YaxisMatrix:
                    size = size + sys.getsizeof(o)
                    for o2 in o:
                            size = size + sys.getsizeof(o2)
            for o in self.u:
                    size = size + sys.getsizeof(o)
            return size
class envt:
    """A grid class that stores the details of the
environment."""
    def __init__(self, Xsource, Ysource, V_y):
```

```
            self.dt = donnees.dt
            self.nx = donnees.quad_x
            self.ny = donnees.quad_y
            self.sautx = donnees.saut_x
            self.sauty = donnees.saut_y
            self.Vy = V_y
            self.Xsource = Xsource
            self.Ysource = Ysource
            self.XaxisMatrix = numpy.zeros((self.ny,self.
nx),dtype = int)
            self.YaxisMatrix = numpy.zeros((self.ny,self.
nx),dtype = int)
            for i in range(self.ny):
                    self.XaxisMatrix[i,:] = numpy.
arange(0,self.nx)
            for i in range(self.nx):
                    self.YaxisMatrix[:,i] = numpy.
arange(0,self.ny)
            self.lamda = numpy.sqrt((donnees.env_D*1.*donnees.
env_tau)/(1+ (self.Vy*self.Vy*donnees.env_tau*1.)/(4.*donnees.
env_D)))
            self.u = numpy.ones((self.ny, self.nx), dtype =
float)
            self.g = self.detRate_grid()
    def detRate_grid(self):
            "Detection-rate grid for each point in the grid
given a position for the agent"
            # Matrix of size [ny,nx] with all the distances
between the agent an any other point of the grid
            module_dist = numpy.sqrt(numpy.power((self.Xsource-
self.XaxisMatrix)*self.sautx, 2) + numpy.power((self.Ysource-
self.YaxisMatrix)*self.sauty, 2))
            # Average concentrations given previous matrix
            conc = donnees.env_R*1./(4.*numpy.pi*donnees.
env_D*module_dist) * numpy.exp(-1.*(-self.Ysource+self.
YaxisMatrix)*self.sauty*self.Vy/(2.*donnees.env_D)) *numpy.
exp(-1.*module_dist/self.lamda)
            conc[self.Ysource,self.Xsource] = 100
            # Detection-rate matrix
            g = 4.*numpy.pi*donnees.env_D*donnees.env_a*conc
            return g
    def poissonCum(self, Npoints, landa):
            increasingNbs = numpy.linspace(0, Npoints-1,num =
Npoints)
            poissonVals = numpy.power(landa,increasingNbs) *
numpy.exp(-landa) *1./factorial(increasingNbs)
            return cumsum(poissonVals)
    def particules(self,Xpos,Ypos):
            Npoisson = 5
            landa = self.g[Ypos,Xpos]*self.dt
            distrib = self.poissonCum(Npoisson, landa)
```

```
            r = numpy.random.uniform(0,1)
            Nparticules = numpy.size(numpy.nonzero(r>distrib))
            return Nparticules
    def plot(self):
            fig = plt.figure()
            ax = fig.add_subplot(111)
            plt.pcolormesh(numpy.log(self.g))
            plt.axis([0,self.nx,0,self.ny])
            fig.savefig('sourceFig')
            plt.show()
```

5.6.2 Spiral-Surge

For the case of spiral-surge, the possible behaviors for the agent were explicitly
described as below:

```
import numpy
import donnees
import sys
class Robot:
    def __init__(self, xPosition, yPosition):
            # position of robot
            self.xPosition = xPosition
            self.yPosition = yPosition
            self.x = self.xPosition
            self.y = self.yPosition
            self.teta = 0.5
            self.dSurge = 0.
            self.up = 0
            self.trans = 0.
            self.horizontal = 0.
    def surge(self):
            self.dSurge = self.dSurge - donnees.saut_y
            self.yPosition = self.yPosition + donnees.saut_y
    def init_spiral(self):
            self.x = self.xPosition
            self.y = self.yPosition
            self.teta = 0.5
    def spiral(self):
            r = (donnees.dGap * self.teta)/(2*numpy.pi)
            self.xPosition = r * numpy.cos(self.teta) + self.x
            self.yPosition = r * numpy.sin(self.teta) + self.y
            self.teta = self.teta + (donnees.saut_x)/r
```

REFERENCES

Baker, T., Willis, M., Haynes, K., and Phelan, P. 1985. A pulsed cloud of sex pheromone elicits
    upwind flight in male moths. *Physiol. Entomol.* 10: 257–265.
Balkovsky, E., and Shraiman, B. I. 2002. Olfactory search at high Reynolds number. *Proc.
    Natl. Acad. Sci. U.S.A.* 99(20): 12589–12593.

Barbieri, C., Cocco, S., and Monasson R. 2011. On the trajectories and performance of info-taxis, an information based greedy search algorithm. *Europhys. Lett.*, 94.

Berg, H. C. (2003). *E. coli in motion*. New York: Springer.

Cardé, R. T., Cardé, A. M., and Girling, R. D. 2012. Observations on the flight paths of the day-flying moth *Virbia lamae* during periods of mate location: Do males have a strategy for contacting the pheromone plume? *J. Anim. Ecol.* 81(1): 268–276.

Collett, T. S., and Collett, M. 2002. Memory use in insect visual navigation. *Nat. Rev. Neurosci.* 3: 542–552.

Consi, T., Grasso, F., Mountain, D., and Atema, J. 1995. Explorations of turbulent odor plumes with an autonomous underwater robot. *Biol. Bull.* 189: 231–232.

Grasso, F. 2001. Invertebrate-inspired sensory-motor systems and autonomous, olfactory-guided exploration. *Biol. Bull.* 200: 160–168.

Grasso, F., Dale, J., Consi, T., Mountain, D., and Atema, J. 1997. Effectiveness of continuous bilateral sampling for robot chemotaxis in a turbulent odor plume: Implications for lobster chemo-orientation. *Biol. Bull.* 193: 215–216.

Hayes, A. T., Martinoli, A., and Goodman, R. M. 2002. Distributed odor source localization. *IEEE Sensors J.* 2(3): 260–271.

Hugues, E., Rochel, O., and Martinez, D. 2003. Navigation strategies for a robot in a turbulent odor plume using bilateral comparison. Presented at International Conference on Advanced Robotics (ICAR), University of Coimbra, Portugal, June 30–July 3.

Jones, C. 1983. On the structure of instantaneous plumes in the atmosphere. *J. Hazard. Mat.* 7: 88–112.

Kaissling, K.-E. 1997. Pheromone-controlled anemotaxis in moths. In *Orientation and communication in Arthropod*, ed. M. Lehler, 343–374. Basel, Switzerland: Birkhäuser Verlag.

Kanzaki, R., Ando, A., Sakurai, T., and Kazawa, T. 2008. Understanding and reconstruction of the mobiligence of insects employing multiscale biological approaches and robotics. *Adv. Robotics* 22(15): 1605–1628.

Kanzaki, R., Arbas, E. A., and Hildebrand, J. G. 1991. Physiology and morphology of descending neurons in pheromone-processing pathways in the male moth *Manduca sexta*. *J. Comp. Physiol. A* 169: 1–14.

Kanzaki, R., Ikeda, A., and Shibuya, T. 1994. Morphological and physiological properties of pheromone-triggered flipflopping descending interneurons of the male silkworm moth, *Bombyx mori*. *J. Comp. Physiol. A* 175: 1–14.

Kraus-Epley, K., and Moore, P. 2002. Bilateral and unilateral antennal lesions alter orientation abilities of the crayfish, *Orconectes rusticus*. *Chem. Senses* 27: 49–55.

Kuwana, Y., Nagasawa, S., Shimoyama, I., and Kanzaki, R. 1999. Synthesis of the pheromone-oriented behaviour of silkworm moths by a mobile robot with moth antennae as pheromone sensors. *Biosensors Bioelectronics* 14:195–202.

Lilienthal, A., and Duckett, T. 2003. Experimental analysis of smelling Braitenberg vehicles. Presented at International Conference on Advanced Robotics (ICAR), University of Coimbra, Portugal, June 30–July 3.

Lochmatter, T. 2010. Bio-inspired and probabilistic algorithms for distributed odor source localization using mobile robots. PhD thesis, EPFL, Lausanne.

Lochmatter, T., Raemy, X., Matthey, L., Indra, S., and Martinez, A. 2008. A comparison of casting and spiraling algorithms for odor source localization in laminar flow. In *Proceedings of the 2008 IEEE International Conference on Robotics and Automation (ICRA 2008)*, 1138–1143. Pasadena, California.

Lytridis, C., Kadar, E. E., and Virk, G. S. 2006. A systematic approach to the problem of odour source localisation. *Autonomous Robots* 20(3): 261–276.

Marques, L., Nunes, U., and de Almeida, A. T. 2006. Particle swarm-based olfactory guided search. *Autonomous Robots* 20(3): 277–287.

Martinez, D. 2007. On the right scent. *Nature* 445: 371–372.

Martinez, D., and Perrinet, L. 2002. Cooperation between vision and olfaction in a koala robot. In *Report on the 2002 Workshop on Neuromorphic Engineering*, Telluride, CO, pp. 51–53.

Martinez, D., Rochel, O., and Hugues, E. 2006. A biomimetic robot for tracking specific odors in turbulent plumes. *Autonomous Robot* 20(3): 185–195.

Masson, J.-B., Bailly-Bechet, M., and Vergassola, M. 2009. Chasing information to search in random environments. *J. Phys. A Math. Theor.* 42, 434009. DOI: 10.1088/1751-8113/42/43/434009.

McMahon, A., Patullo, B., and Macmillan, D. 2005. Exploration in a t-maze by the crayfish cherax destructor suggests bilateral comparison of antennal tactile information. *Biol. Bull.* 208: 183–188.

Moraud, E. M., and Chicca, E. 2011. Toward neuromorphic odor tracking: Perspectives for Space exploration. *Acta Futura* 4: 9–20.

Moraud, E. M., and Martinez, D. 2010. Effectiveness and robustness of robot infotaxis for searching in dilute conditions. *Front. Neurorobotics* 4: 1. DOI: 10.3389/fnbot.2010.00001.

Murlis, J., Elkinton, J. S., and Cardé, R. T. 1992. Odor plumes and how insects use them. *Annu. Rev. Entomol.* 37: 505–532.

Murlis, J., Willis, M. A., and Cardé, R. T. 2000. Spatial and temporal structures of pheromone plumes in fields and forests. *Physiol. Entomol.* 25(3): 211–222.

Olberg, R. M. 1983. Pheromone-triggered flip-flopping interneurons in the ventral nerve cord of the silkworm moth, *Bombyx mori*. *J. Comp. Physiol.* 152: 297–307.

Pang, S., and Farrell, J. 2006. Chemical plume source localization. *IEEE Trans. Syst. Man Cybern. B* 36: 1068–1080.

Pyk, P., Bermúdez i Badia, S., Bernardet, U., Knüsel, P., Carlsson, M., Gu, J., Chanie, E., Hansson, B. S., Pearce, T. C., and Verschure, P. F. M. J. 2006. An artificial moth: Chemical source localization using a robot based neuronal model of moth optomotor anemotactic search. *Autonomous Robots* 20(3): 197–213.

Rains, G. C., Tomberlin, J. K., and Kulasiri, D. 2008. Using insect sniffing devices for detection. *Trends Biotechnol.* 6(6): 289–290.

Rajan, R., Clement, J., and Bhalla, U. 2006. Rats smell in stereo. *Science* 311: 666–670.

Reynolds, A. M., Reynolds, D. R., Smith, A. D., Svensson, G. P., and Löfstedt, C. 2007. Appetitive flight patterns of male *Agrotis segetum* moths over landscape scales. *J. Theor. Biol.* 245(1): 141–149.

Roberts, P., and Webster, D. 2002. Turbulent diffusion. In *Environmental fluid mechanics, theories and applications*, ed. H. Shen, A. Cheng, K.-H. Wang, M. H. Teng, and C. Liu, 7–45. Reston, VA: ASCE Press.

Russell, R. 1999. *Odour detection by mobile robots*, 22. World Scientific Series in Robotics and Intelligent Systems. River Edge, NJ: World Scientific Publishing Co.

Russell, R. A., Bab-Hadiashar, A., Shepherd, R. L., and Wallace, G. G. 2003. A comparison of reactive robot chemotaxis algorithms. *Robotics Autonomous Syst.* 45(2): 83–97.

Vergassola, M., Villermaux, E., and Shraiman, B. I. 2007. Infotaxis as a strategy for searching without gradients. *Nature* 445: 406–409.

Vickers, N. J. 2006. Winging it: Moth flight behavior and responses of olfactory neurons are shaped by pheromone plume dynamics. *Chem Senses* 31: 155–166.

Wehner, R., and Menzel, R. 1990. Do insects have cognitive maps? *Annu. Rev. Neurosci.* 13: 403–414.

Weissburg, M. 2000. The fluid dynamical context of chemosensory behavior. *Biol. Bull.* 198: 188–202.

6 Performance of a Computational Model of the Mammalian Olfactory System

*Simon Benjaminsson, Pawel Herman,
and Anders Lansner*

CONTENTS

We compare a biomimetic model of the early mammalian olfactory system and the olfactory cortex with machine learning methods on odor classification and segmentation problems for input data having biologically plausible characteristics and for polymer sensor recordings. A cortical adaptation-based mechanism for odor segmentation is proposed and context-dependent segmentation is demonstrated in the model. The capability of biomimetic models to integrate several mechanisms in one system may prove highly relevant in addressing complex real-world chemosensory tasks.

6.1   INTRODUCTION

Sensory inputs allow the brain to construct a detailed multimodal map of the surrounding world. Smell is a carrier of critical information about food, predators, and social status, among others. Thus, a wide range of mammalian or insect behaviors rely on the recognition of relevant odor stimuli. In the olfactory system, it involves the identification of a composition of molecules in rich odorant mixtures. Considering ecological aspects of this fundamental sensory task, in particular the complexity of naturally occurring odors blended in the chemically noisy environments and variability in olfactory stimulation conditions, manifested in the flux of perceived quality and intensity, the objective to maintain stable object perception is challenging. In order to construct the percept, the system must detect a chemical stimulus, extract its relevant features from fluctuating odiferous backgrounds, create their robust representations, and match them with odor patterns experienced earlier and stored in memory (Cleland and Linster 2005). The fundamental computational challenges in robust odor recognition beyond simple discrimination, which biological olfaction successfully addresses to instigate complex behaviors, include odor concentration-invariant identification, background elimination, and mixture segmentation (Hopfield 1999). From the perspective of perceptual learning in the olfaction system, the capabilities to memorize, categorize, generalize, and process odor information in a context-dependent fashion underlie its aforementioned functionality.

Despite intensive research efforts and the abundance of experimental material ranging from molecular or electrophysiological evidence to psychophysical and psychological data gathered over the past few decades, there is still a plethora of open questions and debatable hypotheses. We adopt and present a different approach to studying the mammalian olfactory system in this contribution. Namely, we attempt to understand general principles of olfactory information processing and its functional implications by constructing a computational model. The prominent role of olfactory modeling in verifying the existing and presenting novel hypotheses has been well recognized (Cleland and Linster 2005; Pearce 1997). Since we treat the system holistically, our model encompasses the first and second stages of mammalian early olfactory processing along with the first-level olfactory cortical structure, the olfactory (piriform) cortex (OC). More specifically, we synthesize olfactory stimuli patterns and simulate their processing in the olfactory epithelium (OE) in the form of olfactory receptor neuron (ORN) activations. To ensure a satisfactory level of biological plausibility of the model, the distribution of ligand–olfactory receptor (OR) affinities is generated to account for key statistical features of widely reported in vivo primary odor representations. The resulting olfactory codes are then processed in the reduced model of the olfactory bulb (OB), which constitutes an interface between the OE and the OC. In this study, OB computations are handled within the modular structure of glomerular columns, where the ORNs expressing the same OR converge and project to the corresponding group of second-order neurons. The transformation of the primary to the secondary odor representation is defined as an integration of incoming ORN activations with positive (excitatory contribution) and negative (inhibitory contribution) weights to implement a novel interval concentration coding scheme. In consequence, sparse stimulus intensity-dependent olfactory representations are produced at the output of the OB for further processing in the OC, which lies at the heart of our biomimetic approach to the odor recognition problem. The OC model is implemented in the framework of an attractor network with a hierarchical modular architecture.

The proposed three-stage abstract model of key biological mechanisms involved in olfactory information processing operates on static rate-based neural representations, which are believed to convey relevant aspects of information about odor quality and concentration (Olsen et al. 2010). Temporal features of olfactory codes and computations are therefore out of scope of this contribution. We are also interested in the scalability of the model, specifically of the OC due to its involvement in most demanding computations and the biologically plausible dimensionality of odor representations.

The study reported in this chapter is aimed at evaluating the developed abstract model of the mammalian olfactory system in test cases involving key computational tasks in odor object recognition—classification, mixture segmentation, and context-dependent identification of olfactory stimuli. The proposed mechanisms underlying the olfactory function are demonstrated in this regard. The focus here is on biomimetic aspects of these evaluation scenarios, and thus the attributes of perceptual learning and associative memory framework of the piriform cortex model, such as concentration invariance, generalization, and pattern completion, are given special attention. Our holistic modeling approach also provides insight into the impact of early olfactory coding and stimulus representations on the recognition performance. In quantitative terms, we adopt the evaluation criterion as the percentage rate of successful detection of target odor objects.

The formulated problems of odor object recognition, mainly classification and mixture segmentation, constitute a typical set of tasks addressed in machine olfaction. Despite significant differences between neural olfactory representations of natural odors and chemosensory responses to synthetic stimuli, a range of analogies in data processing and pattern analysis have been pointed out (Pearce 1997). In this study, we juxtapose the proposed biomimetic approach with classical methods in machine olfaction. One of the advantages of a biomimetic model implementing several functions in a single-system framework is the possibility to perform more than one simple task, e.g., odor discrimination, in complex scenarios more commonly encountered in natural environments. We touch upon this issue when studying a context-dependent segmentation problem. In a broader perspective, such a biomimetic network model could be combined with models of other modalities, e.g., visual, to address real-world multimodal sensory perception tasks.

The chapter is organized as follows. In the next section our biomimetic model is described in more detail. This is followed by a brief discussion of the conventional pattern recognition techniques commonly applied in machine olfaction and employed in this study for comparative purposes. The subsequent parts of the chapter are devoted to the results of evaluation of the olfactory model in test scenarios—odor object classification, mixture segmentation, and context-dependent recognition. The results are reported on the synthesized early olfactory representations and on the large-scale polymer sensor data set. The chapter concludes with a discussion of the major implications of the presented approach despite some limitations, its applicability to solve real-world odor recognition tasks in machine olfaction, and relevance to research on biological olfaction systems.

6.1.1 BIOMIMETIC MODEL

The three-stage model of the mammalian olfactory system, implementing the early olfactory system, is shown in a schematic drawing in Figure 6.1. It comprises the OE and OB, and a population implementing holistic processing capabilities and corresponding to the OC. Each stage is described in detail below.

6.1.2 OLFACTORY RECEPTOR NEURON (ORN) LAYER

The primary representation of an odor stimulus arising at the ORN level is generally described as a high-dimensional nontopographical and distributed code due to the multitude of ORNs with broad tuning properties (Rospars et al. 2000). The characteristic feature of ORNs is their sensitivity/threshold and an increasing activation as a result of stimulus intensification according to a nonlinear concentration-response relationship. We thus model the activity, $f_{r,i}^{\mathrm{ORN}}$, of an individual ORN unit, i, in terms of its mean frequency response (rate coding) to an odor stimulus based on sigmoidal function

$$f_{r,i}^{\mathrm{ORN}} = 1 - e^{-\gamma_i (g_r - \theta)}$$

where γ_i is a gain parameter responsible for modulation of concentration-frequency characteristics of individual ORNs, Θ is a threshold parameter, and g_r is the conductance of the ORN determined as the product of the maximum conductance, g_{or}, and the degree of occupancy of the corresponding OR, o_r, by an odor stimulus:

$$o_r = 1 - \left(1 + \left(\sum_l \frac{M_l}{K_{D_{l,r}}} \right)^{n_r} \right)^{-1}$$

where n_r is the Hill coefficient for the given receptor type, r.

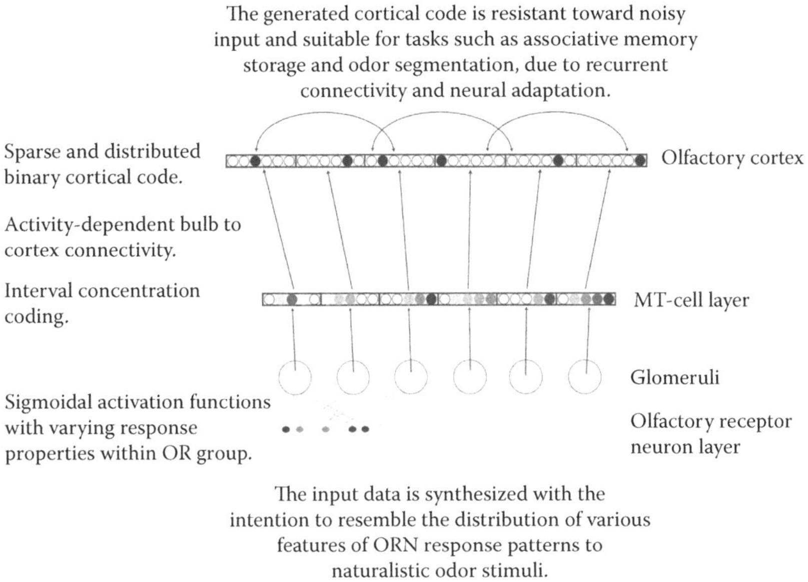

The generated cortical code is resistant toward noisy
input and suitable for tasks such as associative memory
storage and odor segmentation, due to recurrent
connectivity and neural adaptation.

Sparse and distributed
binary cortical code.

Olfactory cortex

Activity-dependent bulb to
cortex connectivity.

Interval concentration
coding.

MT-cell layer

Glomeruli

Sigmoidal activation functions
with varying response
properties within OR group.

Olfactory receptor
neuron layer

The input data is synthesized with the
intention to resemble the distribution of various
features of ORN response patterns to
naturalistic odor stimuli.

FIGURE 6.1   The three-stage olfactory model: OE, OB, and OC. The feedback projection from the OC to the OB is not shown.

In this setup, the gain parameter renders systematically varying response properties of individual ORNs within the same OR family, as suggested by recent experimental evidence (Grosmaitre et al. 2006). The stimulus is defined here as a mixture of ligands l (or a single compound) at concentrations M_i, which interact with the ORs expressed by ORNs according to their affinities. As can be seen, coding of mixtures at the ORN layer level reflects a certain degree of superposition of the component patterns (Rospars et al. 2008). In addition, it should be emphasized that the distribution of OR affinities to ligands, stored in the matrix K_D ($L \times R$, where R is the number of OR groups equivalent to the number of glomeruli and L is the total number of single-compound stimuli that the ORN layer is sensitive to), largely defines the response patterns of the entire ORN layer. Therefore, we aimed to generate the K_D matrix that would encourage preservation of key statistical features of biological olfactory codes, such as ~10 to 20% sparseness level, high degree of overlap between chemical receptive fields of ORs, and a limited concentration domain for each odor occurring in natural environments (Hopfield 1999; Malnic et al. 1999; Rospars et al. 2003), in the resulting synthetic odor representations.

6.1.3   OLFACTORY BULB (OB)

The populations of ORNs expressing the same OR have been found to typically target no more than two spherical structures in the OB called glomeruli, one on each side of the brain (Cleland 2010). This reflects an important feature of early olfactory coding, high convergence—millions of ORNs project to a few thousand glomeruli. The design of our abstract model rests on the simplified assumption that

ORNs expressing the same OR type converge onto one glomerulus and rely on a lower level of convergence, which is functionally associated with heightening the system's sensitivity to stimuli and increasing the signal-to-noise ratio by averaging out uncorrelated noise (Laurent 1999).

The secondary representation of an olfactory stimulus arises in the population of second-order olfactory principal neurons—mitral and tufted (MT) cells that also project to the next stage, the OC. They have single primary dendrites that terminate in one glomerulus each and several secondary dendrites that innervate the respective neighborhood (Mori et al. 1999). Additionally, local inhibitory interneurons, granule and periglomerular cells (Chen and Shepherd 2005), play an important role in shaping the responses of principal neurons (Cleland 2010). In our model, the key aspect of these bulbar computations was an effective handling of stimulus intensity-related information, which led to a novel concept of interval concentration coding in the secondary olfactory representation (Sandström et al. 2009). The hypothesis, which goes beyond a limited view of the OB's role in gain control and noise reduction (Chen and Shepherd 2005) (see also Section 6.5), implies that different MT cells in the same glomerulus cover overlapping concentration intervals, hence having receptive fields with soft boundaries in the stimulus intensity domain. As a result, the level of MT cell activity depends on how similar the stimulus concentration is to the preferred (receptive) range. The suitability and robustness of the proposed coding scheme has been initially demonstrated in the context of machine olfaction in odor recognition tasks involving identification of stimulus intensity (Herman and Lansner 2010). Here, the proposed paradigm is implemented using an abstract local differentiation model, which assumes that MT units receive a pool of inhibitory and excitatory connections from populations of ORNs (Figure 6.2). The weights are then distributed to form overlapping Gaussian integrating windows (parameter K controls the amount of overlap) w of the size N (centered at the origin), to simulate the MT unit's responses (ith unit receiving input from glomerulus r; in the standard setup, $i = 1, \ldots, 25$ and $r = 1, \ldots, 300$, see Section 6.2), $f_{r,i}^{MT}$,

$$f_{r,i}^{MT} = \sum_{j=1}^{N} w\left(\frac{N}{2} - j\right) \cdot \left(f_{r,j+(i-1)N-K}^{ORN} = f_{r,j+iN-K}^{ORN}\right)$$

according to the following equation:

In the simplest case, the window is reduced to one point and the MT response becomes the result of subtraction of activations of the corresponding neighboring ORNs (they are ordered with respect to their sensitivity). Lateral inhibition and competitive learning in the OB are assumed to be among the underlying mechanisms to support such a coding (Laurent 1999; Cleland and Sethupathy 2006).

In subsequent analyses of secondary olfactory representations, the normalized MT responses are used. Such normalization is performed to ensure a constant level of overall activity (one in arbitrary units) in the group of MTs for concentrations exceeding a certain threshold (near the K_D value of interest). It is assumed to result from the negative feedback produced by inhibitory granule cells onto the MT cells. Examples of ORN responses (subset of ORNs expressing the same OR type) and

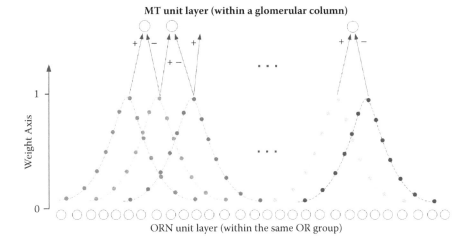

MT unit layer (within a glomerular column)

Weight Axis

ORN unit layer (within the same OR group)

FIGURE 6.2 Conceptual diagram illustrating the structure of connectivity from the ORN array to MT units within the same glomerular column. The ORNs in the input layer are ordered with respect to their sensitivity. The activity of each MT unit is determined as the weighted sum of the activities of the ORNs within two neighboring Gaussian weight windows (one integrates excitatory, i.e., positive, contributions, and the other one inhibitory, i.e., negative, contributions). The weight axis describes the normalized weight strength.

the corresponding MT activity, as well as normalized MT activity, are shown in Figure 6.3.

6.1.4 OLFACTORY CORTEX (OC)

The cortex implements the holistic processing capabilities underlying classification, segmentation, and context dependence.

The recurrent connections spanning the entire cortical population were trained using an online version of the Hebbian-based BCPNN learning rule (Lansner and Holst 1996; Sandberg et al. 2000) where the final activity $\hat{\pi}_{ii'}(t)$ of a unit is set from

$$\frac{dh_{ii'}(t)}{dt} = \beta_{ii'}(t) + \sum_{j}^{N} \log\left(\sum_{j'}^{M_i} w_{ii'jj'}(t)\hat{\pi}_{jj'}(t)\right) - h_{ii'}(t)$$

and

$$\hat{\pi}_{ii'}(t) = \frac{e^{h_{ii'}}}{\sum_{j} e^{h_{ij}}}$$

for a graded normalized activity, or for a binary winner-takes-all setting,

$$\hat{\pi}_{ii'}(t) = \begin{cases} 1 & \text{if } \max_{\forall j}(h_{ij}) = h_{ii'} \\ 0 & \text{else} \end{cases}$$

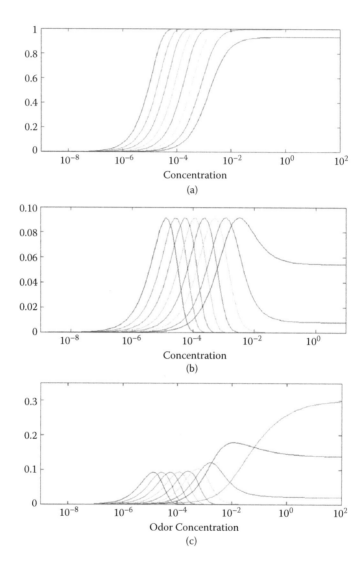

FIGURE 6.3 Examples of dose-response characteristics for (a) ORNs, (b) MTs, and (c) normalized MT activity.

Incremental estimations of activity probabilities and the weights are set from

$$\frac{d\Lambda_{ii'}(t)}{dt} = \alpha(((1-\lambda_0)\hat{\pi}_{ii'}(t)+\lambda_0)-\Lambda_{ii'}(t))$$

$$\frac{d\Lambda_{ii'jj'}(t)}{dt} = \alpha(((1-\lambda_0)\hat{\pi}_{ii'}(t)\hat{\pi}_{jj'}(t)+\lambda_0^2)-\Lambda_{ii'jj'}(t))$$

$$\beta_{ii'}(t) = \log(\Lambda_{ii'}(t))$$

$$w_{ii'jj'}(t) = \frac{\Lambda_{ii'jj'}(t)}{\Lambda_{ii'}(t)\Lambda_{jj'}(t)}$$

Learning variables were set to $\alpha = 0.005$ and $\lambda_0 = 0.0001$ in all experiments.

During a training phase, the response properties of the units in the cortical layer are set up so to be tuned to prototypical patterns in the training set using a competitive learning algorithm, competitive selective learning (CSL) (Ueda and Nakano 1994). This provides an abstract model of the competitive learning assumed to occur in the OB. CSL uses an input pattern x-i., which could be a part of an input odor, to update the position of the closest prototype y by

$$y = y + \varepsilon_{,,} x\text{-}i. - y$$

using a Euclidean distance measure, and where ε determines the amount of change in each iteration and can be gradually decreased during the training phase. To avoid local minima, CSL reinitializes certain prototypes using a selection mechanism according to the equidistortion principle (Ueda and Nakano 1994). Activation of a unit is also determined by the Euclidean distance between a sample and prototype and a winner-takes-all mechanism, where the unit with the most similar prototype is activated and all other units in the module are turned off.

The cortical population implements neural adaptation, which makes the cortical response triggered by an odor gradually decrease. This allows for segmenting mixtures into their components, by letting the cortical response switch between found components. The pattern completion capability enhances this functionality, as only a partial stimulus is enough to make the cortical activity state move into one of the learned odors. In detail, the neural response for a unit with neural adaptation is determined by

$$h_{ii'}(t) = h_{ii'}(t) - a_{ii'}(t)$$

$$\frac{da_{ii'}(t)}{dt} = \begin{cases} (\lambda_{ad} - a_{ii'}(t))\tau_{ad} & \text{if unit active} \\ -a_{ii'}(t)k_{ad}\tau_{ad} & \text{if unit inactive} \end{cases}$$

where is an adaptation amplitude determining maximum adaptation change, τ_{ad} is a time constant determining the speed of the adaptation and its rise time, and k_{ad} is a constant deciding the ratio between rise and decay times. In the simulation experiments, if nothing else is stated, the variables were set to $\lambda_{ad} = 4$, $\tau_{ad} = 0.1$, and $k_{ad} = 5$.

The connectivity between the OB and the cortex is generated (self-organized) such that it is dependent on the OB in such a way that parts that display similar activity are projected to the same region of the cortex (Lansner et al. 2009). This structural plasticity is performed by first estimating the mutual information between each glomerulus, grouping them according to these values, and setting up the connectivity from each group of glomeruli to a module in the cortex. In detail, the mutual information between a pair of glomeruli is calculated from

$$I(X,Y) = \sum_i \sum_j p_{ij} \log \frac{p_{ij}}{p_i p_j}$$

Here, i and j are the indices for the units in each glomeruli, and the unit activities are used to estimate the probabilities online as

$$\frac{dp_i(t)}{dt} = p_i(t) + \alpha_m (A_i(t) - p_i(t))$$

and

$$\frac{dp_{ij}(t)}{dt} = \alpha_m (A_i(t) A_j(t) - p_{ij}(t))$$

$A_i(t)$ is the unit activity and $\alpha_m = 0.003$ is used in all experiments.

With the joint entropy calculated as

$$J(X,Y) = -\sum_i \sum_j p_{ij} \log p_{ij}$$

the mutual information can be transformed into a distance measure (Kraskov and Grassberger 2009):

$$D(X,Y) = 1 - \frac{I(X,Y)}{J(X,Y)}$$

By utilizing classical multidimensional scaling (Young 1985), a new map is created that is incrementally changed over time to fulfill these calculated distance relations between each pair of glomeruli. The glomeruli are clustered again using CSL in this map into the same number of clusters as cortical modules. These clusters decide to which cortical module each glomerulus was connected.

6.1.5 CLASSICAL METHODS

The proposed biomimetic model of a mammalian olfactory system has been evaluated here in a comparative framework including classical pattern recognition methods commonly applied in the area of machine olfaction (Gutierrez-Osuna 2002; Hines et al. 2003). Both supervised and unsupervised approaches are represented in this study. In particular, linear support vector machine (SVM) and Gaussian-kernel SVM (SVM$_{Gauss}$) classifiers (Schölkopf et al. 1998), the k-nearest-neighbor (kNN) algorithm (Cover and Hart 1967), and partial least-squares (PLS) discrimination analysis (Wold 1985) have been applied to the problem of odor identification where the target labels were available and utilized during training. In scenarios without or with limited a priori information about odor class assignment of input stimuli, an unsupervised approach has been adopted in the form of clustering. Here, the fuzzy c-means (FCM) algorithm (Bezdek 1981) was used, as it has been demonstrated as a suitable approach to machine olfaction problems (Yea et al. 1994).

A multiclass version of this SVM was developed by combining one versus rest classifiers. The output of these individual classifiers was exploited to estimate a posterior class probability by a sigmoid function (Platt 2000). Then the resulting pairwise probabilities were coupled to obtain an estimate of class probabilities in a multiclass context using a generalized Bradley-Terry model (Huang et al. 2006). This classification system was thus adapted to perform segmentation by thresholding the output class probabilities. The parameters of a linear SVM and an SVM with homoskedastic Gaussian kernel, applied in this work, were selected in the course of a holdout validation process.

PLS was utilized in this evaluation as a robust representative of multivariate linear models in scenarios involving high-dimensional data with relatively few observations hence liable to overfitting. The method is aimed at identifying a few underlying factors that account for the gross variation in the dependent variable, which here, in the context of discriminative analysis, are the classification labels. We employed the nonlinear iterative PLS algorithm, which linearizes models that are nonlinear in the parameters and handles overdetermined regression problems (Wold 1966).

In an unsupervised scenario, a training set of odor representations was clustered using the FCM algorithm without providing any information about their class assignments (odor identity label) except the total number of odors. Due to commonly reported weaknesses of FCM when dealing with high-dimensional data (Winkler et al. 2011), an initialization procedure was devised to provide meaningful initial cluster positions. To this end, one representative sample from each odor class has been designated as an initial cluster seed. The recall phase consisted in translating the membership values of incoming samples, obtained based on their distance from the cluster centers, to the posterior probabilities. It should be mentioned at this point that both Euclidean and correlation-based distance metrics have been applied in this study. In the classification task, the maximum probability determined the class assignment, whereas in the segmentation scenario, the respective number of top probabilities (the number of mixture components) indicated multiple class assignments (single ligands reflected in the odor mixture).

6.2 DATA DESCRIPTION

6.2.1 SYNTHETIC DATA

Synthetic olfactory data utilized in this work have been generated in the first two stages of the odor information processing pipeline in the biomimetic model, namely, ORN and OB layers. The normalized MT representations (see Section 6.1.1), referred to as *mtF*, have served as input data in classification and segmentation scenarios addressed with the use of classical pattern recognition methods and with our abstract olfactory system model.

A set of different data dimensionality configurations have been tested in this evaluation. In particular, the size of the ORN array (the number of ORN units per OR group) and OB layers (the number of glomeruli corresponding to the number of OR groups) has been extensively varied, whereas the number of MT units per glomerulus has been kept within a narrow range of 10 to 30 in accordance with biological data (Royet et al. 1998). In the most representative setup, which has served to produce the majority of results demonstrated in this chapter, 7500 (for the size of the weight window, reduced to one sample; see Section 6.1.1) or 300,000 ORNs are set to project to 300 glomeruli, each of which comprises 24 MT units. The resulting MT representations are 7200-dimensional. In this model configuration, the activation patterns have been simulated for 50 single ligands in total at around 30 to 50 concentration levels uniformly distributed over the log concentration domain assigned individually to each odor object (see Section 6.1.1). In addition, the olfactory data for 20 two-ligand (binary) and 10 three-, four- and five-ligand mixtures have been generated.

The MT layer activation as a result of stimulating the abstract model of an ORN array with five odor patterns at varying concentrations is illustrated in Figure 6.4. The concentration axis is expressed in logarithmic scale reflecting a linear increase of stimuli intensity in log units (ramp pattern of the stimulus concentration sweep). As can be noticed, the code is sparse and distributed. In fact, the level of sparseness (the percentage of active units) depends on the concentration range (Figure 6.5), which can be well understood upon examination of the response patterns at a finer scale, i.e., for subsets of units (see Figure 6.6). As can be observed, with the increasing stimulus concentration more ORN units are recruited, which amounts to saturating *ornF* representations. On the other hand, only a subset of MT units is active at a time for a given concentration level in the spirit of the proposed interval concentration coding (Sandström et al. 2009).

We have also examined the correlation structure of the generated *mtF* and *ornF* codes. It has direct implications for the odor recognition and segmentation performance of the OC model. Figure 6.7 illustrates the pairwise distances between *mtF* and *ornF* representations of five odors at a wide range of concentrations. The distance is calculated as a complement of the correlation coefficient to unity.

Upon examination of Figure 6.7, it appears that *mtF* code is more decorrelated than *ornF*. In addition, the similarity of the *mtF* representations of the same odor identity but of distinct intensities strongly depends on the actual differences in concentration levels. This effect is weaker for *ornF* codes, where the band of ligand

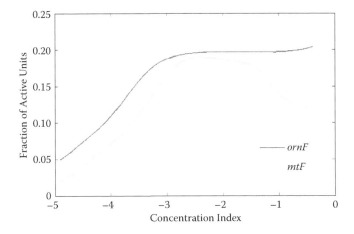

FIGURE 6.4   The activation of MT layer units due to the stimulation with five odor patterns at increasing concentrations. The concentration axis in each of the five horizontal panels is independent (expressed in log units).

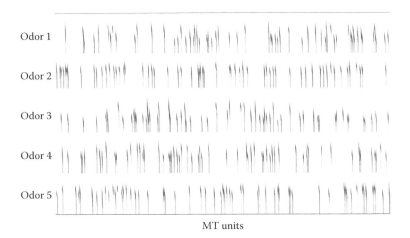

FIGURE 6.5   Sparseness of *ornF* and *mtF* codes for varying concentrations of one of the input odor patterns. The log concentration axis corresponds to the concentration domain of the input odor.

self-similarity is wider (blue diagonal). This observation is a manifestation of the concentration variant nature of activations of the MT units.

Finally, the relationship between early olfactory codes of single odors and their mixture stimuli has been thoroughly examined. The relevant question to address is about the extent to which a mixture representation reflects response patterns characteristic to its individual components. The outcome bears particular relevance to the odor segmentation task performed by the OC model.

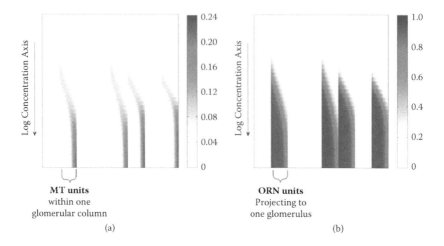

FIGURE 6.6   The activation of a subset of ORN groups (a) and the corresponding MT units (b) for odor stimuli at 45 concentrations spanning the dynamic range of the selected groups. The interval concentration coding concept is manifested in the MT responses.

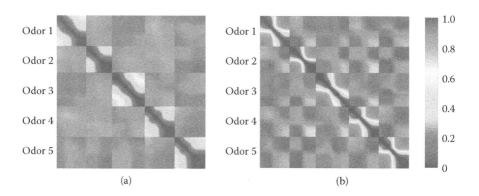

FIGURE 6.7   **(See color insert.)** Distance matrix for (a) *ornF* and (b) *mtF* representations of five odors at different concentrations. The distance is expressed in the interval [0, 1] as a quantity complementing the correlation coefficient to unity.

Figure 6.8 shows two matrices, for *ornF* and *mtF*, depicting Euclidean pairwise distances between olfactory codes corresponding to single odors and their binary mixture stimuli. Analogously, as in the previous case, the order of matrix entries within each odor group is sorted with respect to concentration. The similarity between the mixture and its corresponding single components clearly manifests itself in both matrices. This similarity, also between different mixtures, is again more diffused over a wider range of concentrations for the *ornF* code. Although it facilitates concentration invariant recognition, the risk of spurious recognition is higher.

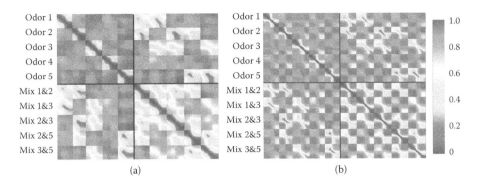

(a)   (b)

FIGURE 6.8 **(See color insert.)** Correlation-based distance matrices for (a) *ornF* and (b) *mtF* representations of five single components and their five binary mixtures (1&2, 1&3, 2&3, 2&5, and 3&5). The horizontal and vertical black solid lines separate the areas corresponding to single odor and binary mixtures.

6.2.2 CHEMOSENSORY DATA FROM A LARGE-SCALE ARRAY OF POLYMER SENSORS

The performance of the proposed biomimetic model has also been benchmarked on an experimental chemosensory data set obtained from a large-scale array of polymer sensors (see Chapter 3). The high-dimensional (16,384 sensors distributed on four equally sized boards) sensory readings were generated for three single compounds such as butanone, ethanol, and acetic acid, as well as for the binary mixture of the first two—butanone and ethanol at varying concentrations.

Prior to the use of the polymer data set in classification and segmentation scenarios, the sensor readings were preprocessed and relevant signal features were extracted. A filtering procedure was employed to remove high-peak noise components and high-frequency noise. Then the data were corrected for a linear drift. An important factor in this filtering process was implicit referencing of the data to average values over a time window in the premeasurement period rather than taking values of a single frame. As a result, we have obtained the time series representative for sensor responses to chemical stimuli.

In order to perform classification and segmentation on the polymer data, the spatiotemporal chemosensory responses were represented first in the feature space. In particular, the features were extracted independently from each time series as the mean over their stationary component, usually over the span of around 100 frames a 16,384-dimensional representation of the array's response to every odor stimulus was obtained.

We have initially examined distinctive odor patterns manifested in the high-dimensional feature vectors by performing principal component analysis (PCA). This has allowed us to illustrate the data in a two-dimensional principal component (PC) space (Figure 6.9), where the major part of variance (~75%) is captured.

The high-dimensional polymer data were further processed with the use of the biomimetic algorithm described above to simulate neural feature extraction at the OB

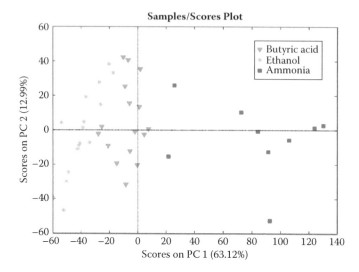

FIGURE 6.9 PCA of the high-dimensional feature representation of the chemosensory array to three compounds at a range of concentrations.

level. In short, correlated activity of individual sensors across the entire data set was quantified with mutual information and served as a basis to build a pairwise sensor distance matrix. This matrix was in turn analyzed with the multidimensional scaling algorithm, which facilitated clustering of the 16,384 sensors and provided a lower-dimensional representation of the original polymer data. The convergence from 16,384- into 192-dimensional space mimicked the biological convergence of ORN projections to OB and resulted in the MT-like code. This representation has been used in classification and segmentation tasks.

6.3 RESULTS ON SYNTHETIC DATA

6.3.1 CLASSIFICATION OF ODOR IDENTITY

In the classification task the aim is to recognize the identity of an input odor (single ligand) object irrespective of its concentration based on previously acquired knowledge about the stimulus space. Here we have examined different scenarios depending on the amount of available data for the olfactory system in the training phase. These cases were parameterized with the fraction of the complete synthetic data collection consisting of odor stimuli at all concentrations (ranging from 30 to 50 levels per odor; see Section 6.2). Training data were uniformly distributed over the entire concentration domain for each odor. Four sets of results are presented in Figure 6.10 for the following proportions of training data: 5, 10, 25, and 50%. The 5% level corresponds roughly to one middle-concentration sample per odor. In the simulations reported here, apart from the *mtF* representations fed into the OC model from the OB, we have also tested a scenario with the omission of the OB model, where *ornF* representations are directly provided to the OC from the ORN layer. This

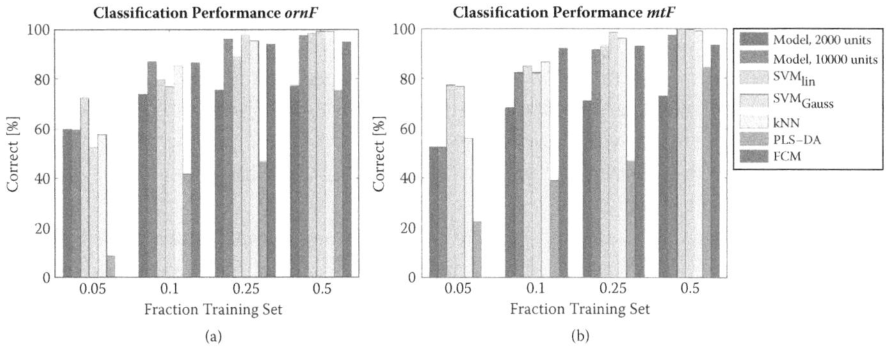

FIGURE 6.10 Classification results for different sizes of the training set for (a) *ornF* and (b) *mtF* representations.

examination has been carried out with a view to making a comprehensive comparison between the biomimetic platform and the classical machine olfaction techniques (see Section 6.1.5) in terms of classification performance evaluated as the classification accuracy (CA).

The classification results obtained using the biomimetic model have been benchmarked against the performance of the conventional pattern recognition techniques considered in this evaluation. The biomimetic model has been run for two different sizes of the cortex, 2000 units or 10,000 units divided into 10 modules. The collective results are depicted in Figure 6.10 for different amounts of training data. Table 6.1 presents the mean CAs along with the standard deviations obtained in 10-trial classification tests when the training set contained 50% instances of the total number of sample stimuli. As can be observed, for both *ornF* and *mtF* representations the performance of all the classification approaches steadily increases with the growing size of training data except for the FCM-based method. This clustering algorithm exploited here as a semisupervised discrimination scheme exhibits comparable performance independently of the training data size provided that each odor cluster in the feature space is seeded with a representative sample (within a middle-concentration range). Although the evaluation outcome points toward the effectiveness of the FCM iterative algorithm in finding the underlying structure of the data, it is particularly sensitive to the metric type in the feature space. In this study, we

TABLE 6.1
Classification Results Using 50% of Training Data

				Classification Methods			
CA [%]	Model (2000 units)	Model (10,000 units)	SVM_{lin}	SVM_{Gauss}	kNN	PLS-DA	FCM
ornF	77.3 ± 1	97.3 ± 0.5	98.1 ± 1.2	99.0 ± 0.9	99.1 ± 0.8	75.3 ± 2.1	94.8 ± 1.3
mtF	72.8 ± 1	97.3 ± 0.5	99.8 ± 0.2	99.6 ± 0.2	99.0 ± 0.4	84.5 ± 1.0	93.3 ± 0.9

have decided to employ the correlation-based distance in the definition of the FCM cost function, following unsuccessful attempts with the standard Euclidean metric. Evaluation of the FCM trained on the 5% level training set has been omitted here, as it boils down to iterating the same data points as those used in the initialization (or in the close neighborhood) without causing any effective update of the cluster centers. In our view, this would not provide an adequate or representative result in the assessment of the FCM approach.

For the biomimetic model with 2000 units, there is a slower increase in performance as the training set size becomes larger compared to the other methods. This type of saturation, which is not seen for the model with 10,000 units, shows the importance of having a large enough cortex. As 2000 is still quite a large number compared to the training set sizes, it hints at the importance of a cortical code that is well adapted to the input data characteristics.

Since SVMs have recently received considerable attention in machine olfaction (Distante et al. 2003), they are taken into account in this study. Although the performance of both SVM classifiers, SVM_{lin} and SVM_{Gauss}, is comparable, the parameter selection procedure is considerably more demanding for the latter approach. It also requires a much higher number of support vectors than SVM_{lin}. For the results reported in Table 6.1 (both $ornF$ and mtF sets containing around 1000 samples), the average number of support vectors used with SVM_{Gauss} is 280, whereas for SVM_{lin} it is 50. The kNN classifier has delivered equally robust performance at the lowest computational costs and without any parameter tuning since $k = 1$ (the number of neighbors) has been fixed in all simulations. It should be realized, however, that even at the testing phase all labeled data points are stored, which can render this approach inefficient for larger data sets.

The PLS-DA approach represents linear methods besides SVM_{lin} in this study. It has proven effective in the analysis of chemosensory data with linear dose-response characteristics (Gutierrez-Osuna 2002; Hines et al. 2003). Here, when evaluated on biomimetic odor representations, both $ornF$ and mtF, it has delivered the lowest performance. More importantly, the reported level of classification accuracy has been obtained for very high numbers of latent variables, suggesting a poor capacity to fit an adequate linear PLS model.

When comparing the results of the odor classification with the use of classical machine olfaction methodology and the proposed biomimetic model, a superior performance of the latter approaches, except PLS-DA, is evident for a smaller cortex size, while the results are similar when compared to a model with a larger number of cortical units. As mentioned earlier, the model has not been devised with the sole aim of maximizing the recognition performance. Other assets of the biologically inspired methodology, especially in the context of biological conditions prevailing in natural environments, should also be accounted for. This aspect of comparative assessment of the odor recognition approaches is elaborated on in Section 6.5.

6.3.2 SEGMENTATION OF ODOR MIXTURES

Segmentation of odor mixtures amounts to the identification of their single components within a certain range of their common concentration domains based on

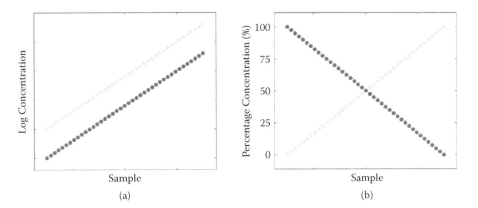

FIGURE 6.11 Conceptual illustration of concentration-dependent mixing scenarios for odor pairs (samples of odor 1 in light gray, odor 2 in dark gray). Each mixture sample is the result of combining two odors at specific concentrations. In the constant concentration ratio setup (a), the log concentrations of both components change proportionally, and in the complementary mixing condition (b), a weak odor 1 stimulus is combined with a strong odor 2 stimulus at one end of the sample axis, and vice versa at the other end.

previously acquired knowledge about the single-ligand stimulus space (optional training sets for the sweeping concentration and single-concentration conditions are the same as in the classification task). The detection performance, quantified as the percentage of successfully segmented mixture samples (i.e., all basic components must be identified), strongly depends on components' intensities within a mixture. This effect has been examined here in two odor mixing scenarios. The first one assumes a constant concentration ratio between the components (constant difference in log concentrations for each mixture regardless of the absolute level; see Figure 6.11a), which could reflect the composition of complex odors present in natural environments. In the other scenario, the sum of the percentage concentration contributions of each component with respect to arbitrary individual reference levels (e.g., concentration level at which the magnitude of the ORN layer activity for the given odor reaches its maximal value) is approximately constant, resulting in complementary mixing proportions across the stimulus intensity axis (Figure 6.11b). This setup has been often employed in psychophysical studies with the use of synthesized odor mixtures (e.g., Muniz et al. 2003). In the evaluation of segmentation performance, only the *mtF* representations have been used.

6.3.2.1 Segmentation of Binary Mixtures in the Biomimetic Model

To evaluate model segmentation performance, we have mixed combinations of two odors using aforementioned mixing scenarios. The model is first trained on a number of single-ligand input patterns and never with the mixture. During testing of segmentation the system is exposed to a mixture of two odors according to the mixing scenario. The task is to identify all the components of the mixture.

As the mixture information progresses to the cortical layer, pattern completion underlies its dynamically stabilization in a trained attractor state corresponding to

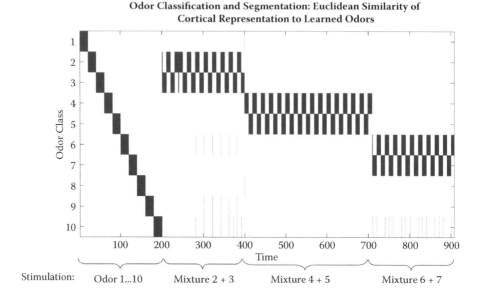

Odor Classification and Segmentation: Euclidean Similarity of Cortical Representation to Learned Odors

FIGURE 6.12   Odor segmentation. The network has been trained on 10 single ligands. Its response to three different double-ligand mixtures demonstrates successful segmentation.

a single odor. Once in a stable state, adaptation causes the activation level to decay, hence allowing for a switch to another attractor state corresponding to another single odor. As neuron adaptation has a rise time, the cortical representation can oscillate between the representations for the different mixture compounds.

Figure 6.12 illustrates the process for a trained system of 10 odors. It shows the Euclidean similarity between the current activity in the cortical population and the activity received when it is exposed to the trained odors. Initially it was subjected to the trained mixtures to show that their cortical representations were separated and clearly defined. Three different mixtures were shown, and the cortical activity was switching between the components of the mixture. To make an easy readout, we have only allowed the system to switch every 20th time step. A successful segmentation is claimed when the cortical population switches between all the components of a mixture and no others.

To evaluate the performance, we have trained a network with a cortical population of 1500 units on the previously described single-odor data set of 50 odors for a total of 2044 items (see Section 6.2). Testing has been carried out on 20 different binary mixtures of varying concentration combinations given the two mixing rules, for a total of 1944 items for each mixing rule. A strong adaptation time constant ($\tau = 0.5$), in combination with a fast rise time for the units depressed by a particular odor, has limited the detection to a maximum of two odor components. The segmentation results for all mixtures are 70% for constant concentration mixing and 13% for complementary mixing proportions. The distributions of the segmentation performance over concentrations are shown in Figure 6.13.

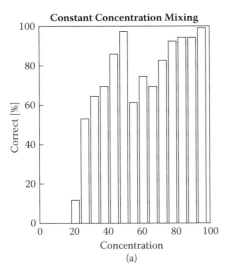
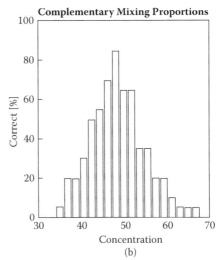

FIGURE 6.13 Segmentation performance distributions for 919 odor mixture stimuli using two mixing rules: (a) constant concentration ratio mixing displays correct segmentation for medium to high concentrations, and (b) mixtures with complementary mixing proportions have a region around medium concentrations where segmentation is successful.

If we look more carefully at the input mixtures, the origins of successful or unsuccessful can be studied. Figure 6.14 shows the Euclidean distances between the representations of a mixture at varying concentrations, its components, and a single-ligand odor that is not part of the mixture. As can be seen, the distance relationships are informative about segmentation performance since the response properties of the cortical units have been set to rely on Euclidean metrics.

The number of learned cortical representations for single odors affects the segmentation performance. Figure 6.15a shows performance when tested on the same constant concentration ratio mixtures as before, but with a varying cortex size. The results seem to saturate at around 1000 to 1500 units in the cortical population, which corresponded to two to three learned cortical representations for all concentrations of each single odor. The more odors that need to be memorized, the larger the cortex needs to be. Figure 6.15b shows a scaling performance of a network consisting only of the cortical population with $6.6 \cdot 10^7$ and its recurrent connections up to 262,144 cores on the JUGENE machine (Benjaminsson and Lansner 2011). The linear speedup shows that it is sufficient to increase the number of cores in order to run networks that in cortical units exceed the sizes of mammalian olfactory cortices. This ensures performance of large networks in real-time applications.

6.3.2.2 Segmentation of Binary Mixtures Using SVM

Segmentation of binary mixtures has also been approached with machine learning methodology. In particular, SVM adopted to multiclass recognition problems (see Section 6.1.5) has been employed to study the distribution of segmentation

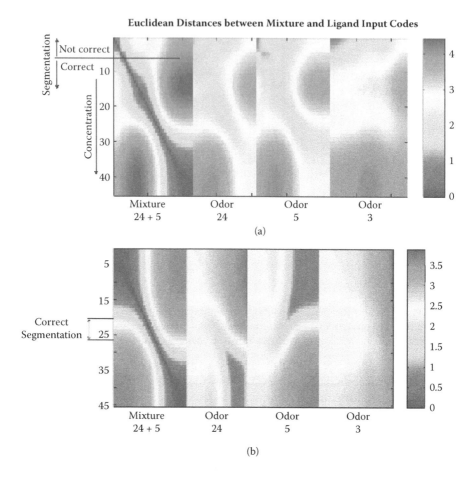

Euclidean Distances between Mixture and Ligand Input Codes

FIGURE 6.14 **(See color insert.)** Euclidean distances between representations of a mixture of two odors, its components, and a single-ligand odor that is not part of the mixture at various concentrations for two different mixture schemes: (a) constant concentration ratio and (b) complementary proportion mixing. Also shown is the segmentation result. Segmentation is successful when the Euclidean distances between odor components and the mixture are low at the same time as distances are high between the mixture and odors not present in the mixture.

accuracy over the span of the concentration parameter in two mixing scenarios (cf. Figure 6.11), analogously to the evaluation of the biomimetic model reported earlier. The results shown in Figure 6.16 display trends similar to those illustrated in Figure 6.13 for the model. It should be noted, however, that SVM allows for the correct segmentation at lower values of concentration parameter than the model for both types of odor mixtures. This has particular impact on the overall segmentation accuracy, averaged over the entire span of the concentration parameter, for mixtures with complementary mixing proportions since the performance is mirrored from the lower to the higher end on the parameter axis (Figure 6.16b; see the caption for the definitions of the concentration parameter).

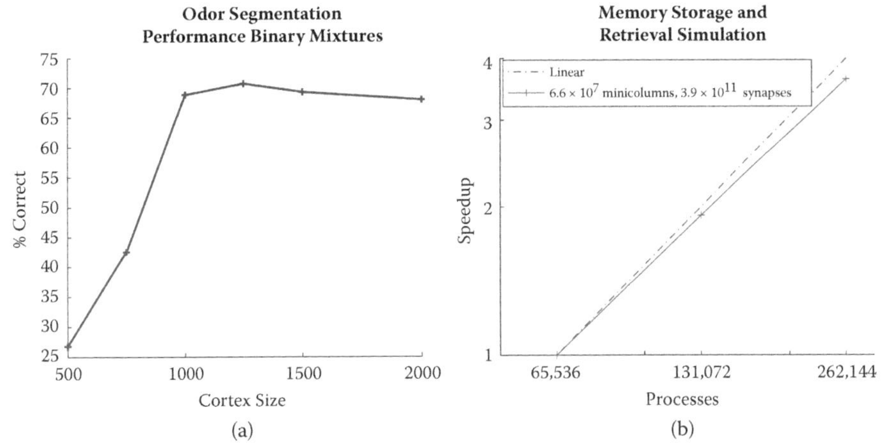

FIGURE 6.15 Scaling performances. (a) Relation between cortex size and performance. Trained on sweeping concentrations of 50 odors and tested on 2044 mixtures. To handle a larger number of odors, the cortex can be scaled up. (b) Linear scaling performance of a network consisting of $6.6 \cdot 10^7$ units and run up to 262,144 cores ensures performance of large networks in real-time applications. (Reproduced from Benjaminsson, S., and Lansner, A., Extreme Scaling of Brain Simulations, Technical Report-FZJ-JSC-IB-2011-02, 7–10, Forschungszentrum Jülich, 2011.)

Figure 6.17 illustrates the comparative set of the overall segmentation results reported with SVM_{lin} and SVM_{Gauss} for two types of odor mixing when the training set with single-component odorants was subsampled, analogously to the evaluation of the classification performance discussed in the preceding section. As expected, the performance steadily increases with a growing size of the training set in both cases (cf. Figure 6.10). The segmentation results obtained with the full training set are collected in Table 6.2. It is interesting to note that SVM_{Gauss} outperforms SVM_{lin} when segmenting mixtures composed of ligands at fixed-ratio concentrations, whereas for mixtures with complementary proportions of the components it is SVM_{lin} that is superior. It should be mentioned here that the computational cost of tuning and evaluating the SVM classifier set is higher for SVM_{Gauss} due to a stronger sensitivity to parameters and a larger number of support vectors, similarly as in the classification setup (see Section 6.3.1).

Both SVMs have delivered considerably higher segmentation rates than the biomimetic model. This is particularly evident in simulations involving mixtures composed of single ligands in complementing proportions. These differences in performance levels are addressed in more detail in Section 6.5.

6.3.2.3 Multiple-Component Mixtures

The system trained on 50 different odors at medium concentration and represented across 7200 MT cells was exposed to 50 different mixture stimuli composed of two, three, four, and five components from the training set, also at medium concentration. The model's performance on mixture component separation has been compared in this study with an SVM_{lin} as a representative of standard machine learning methods.

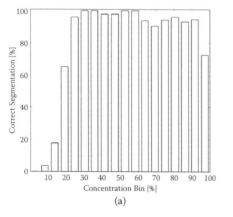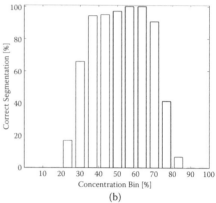

FIGURE 6.16   The distributions of segmentation performance obtained with SVM$_{lin}$ for 919 odor mixture stimuli using two mixing rules: (a) constant concentration ratio mixing displays correct segmentation for medium to high concentrations, and (b) mixtures with complementary mixing proportions. The concentration parameter, (a) log concentration of two components, or (b) ratio of log concentrations of two components, is binned.

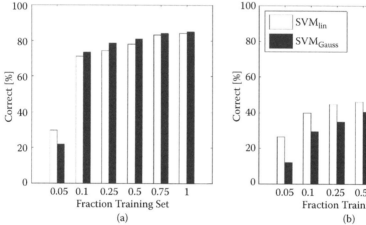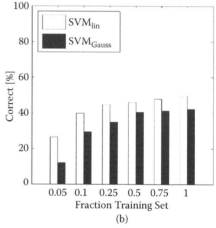

FIGURE 6.17   Segmentation results using SVM on training sets of varying sizes for (a) constant concentration ratio mixing and (b) mixtures with complementary mixing proportions.

Figure 6.18 displays the segmentation accuracy as a function of the number of components in the mixture for the cortical model containing 500 units and the SVM. A performance drop is seen as the number of components is increased. For these more difficult cases, the components detected by the model are part of the mixture but not all of the components are found (not shown in Figure 6.18). The characteristics of the perfomance curves, especially for the cortical model, share resemblance to performance curves reported in human behavioral studies (Laing and Francis 1989). To some extent this is due to the increase of difficulty of the task determined by data characteristics as the number of mixture components is increased (as seen from the

TABLE 6.2
Segmentation Results

[%]	Segmentation Methods		
Mixing Model	**Model (1500 units)**	**SVM_{lin}**	**SVM_{Gauss}**
Constant concentration ratio of individual components	70.0	84.3	85.1
Complementary mixing with varying proportions of individual components	13.1	49.5	42.3

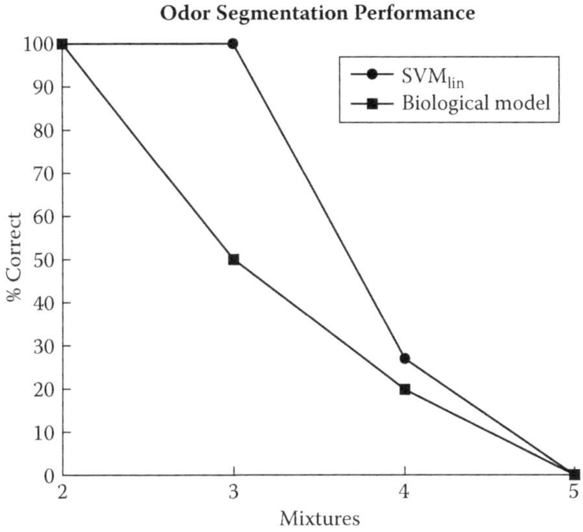

FIGURE 6.18 Segmentation of mixtures composed of two to five components trained and tested on one middle-concentration level. The increased difficulty of the task as the number of mixture components is increased can be observed in both methods. The model shows greater resemblance to human behavioral data. (From Laing, D., and Francis, G., *Physiology & Behavior*, 46(5), 809–814, 1989.)

SVM result). On the other hand, the higher difficulty level manifested in the results on the experimental data (Laing and Francis 1989) and in the model's performance could be due to the specific segmentation mechanism.

6.3.3 Context-Dependent Processing of Odor Stimuli

The model setup for context-dependent processing engaged in particular the feedback projection from the OC to the OB. This projection was trained once the bulb to cortex projection had stabilized. It was restricted to the negative component of the resulting projection since the corresponding synaptic connections in reality are inhibitory, from cortical pyramidal cells to granule cells in the bulb. These connections

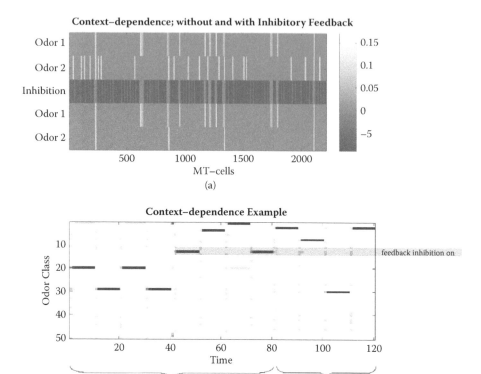

FIGURE 6.19 Context dependence. (a) From the top, OB activity for example odors 1 and 2, the trained inhibitory connections from OC to OB for odor 1, and OB activity for odors 1 and 2 when the feedback projection is turned on. The feedback projection results in no change in the OB activity for odor 1 but strongly inhibits odor 2. (b) A segmentation task of a mixture including odor 13. Odor 13 is detected in the mixture only when the feedback projection trained on it is turned on. With the feedback projection still on, odor 13 is not found when the system is subjected to the mixture that does not contain it.

are presumably also plastic in reality since they contact spines on the granule cells. Here, these inhibitory connections act as a filter for the activity in the bulb, and this filter is tuned by the activity in the cortex or some other type of top-down projection.

Figure 6.19a displays the functionality of the mechanism. The activities of odors 1 and 2 in the OB are shown in the top part of the Figure 6.19. The inhibition from the projection is shown in the middle for a case where it was trained on odor 1. The activities for odors 1 and 2 are in the lower part shown again, with the inhibitory projection turned on. For odor 1, this resulted in no change in activity, while odor 2 was strongly inhibited.

Figure 6.19b shows the functionality in an instance of a mixture segmentation task. Here, a mixture of five components, of which one is odor 13, was presented to the system. Initially the system segmented two components, none of which were odor 13. At $t = 40$, feedback from the cortex was turned on, with the unit trained on odor 13 manually activated, upon which the system found odor 13. It was then subjected to another mixture not containing the odor to show that it did not get activated.

6.4 RESULTS ON POLYMER SENSOR DATA

Similar classification and segmentation tasks have also been performed on the high-dimensional polymer sensor data. Since only one binary mixture was available, the segmentation evaluation has been limited to one simulation. The OC model was trained on trials involving three pure analytes, buthanol, ethanol, and acetic acid, at different concentrations. The testing phase was accomplished using 25 trials of the mixture of ethanol and butanone at three concentration levels (10 and 20 sccm, 20 and 20 sccm, 20 and 10 sccm, respectively). The polymer data representations exploited in these analyses were obtained from the model of self-organizing projections from the ORN to MT layer, as discussed in Section 6.2.2. In consequence, the data dimensionality was reduced from the original size of 16,384 to 192.

As far as the classification of the polymer data is concerned, three specific tasks were devised as follows:

Set 1: For each ligand, trials were divided into three categories based on stimulus concentration; the training set was composed of 50% of trials from each category, and the remaining ones belonged to the test set.

Set 2: The training set was formed out of all trials from the lowest- and medium-concentration range for each ligand. The test set was composed of the remaining trials representing the highest-concentration stimulus. This setup allowed for validation of generalization capacity.

Set 3: The training and test set consisted of trials representing alternate concentration categories, i.e., for six concentration groups, those numbered 1, 3, and 5 formed the training set, and the remaining groups, 2, 4, and 6, the test set.

The results obtained with the reduced OC model have been compared here with the outcome of SVM classification and can be seen in Table 6.3.

The results of classification into one of the three classes of single analytes demonstrate a high level of data separability. They are comparable to the classification accuracy rates obtained with a conventional multiclass SVM_{lin}. These results also reflect the outcome of polymer data characterization reported earlier in Section 6.2.2, where a rather limited size of overlap between clusters of sensor data corresponding to individual analytes can be observed.

With regard to segmentation performance, the biomimetic approach has proven to be on a par with the SVM-based method. It should be noted, however, that achieving 60% level of SVM performance requires an additional assumption about the number of odor components to be identified in the mixture. In other SVM evaluations reported earlier in this chapter, thresholding target class probabilities provide a sufficient measure to identify the correct number of odor components without any prior assumptions.

TABLE 6.3

Classification and Segmentation Results on the Polymer Sensor Data

Classification Tasks	Dimensionality	Classification Accuracy [%]	
		OC Model	SVM$_{lin}$
Set 1	192	86	84
Set 2	192	89	89
Set 3	192	88	90
		Segmentation Accuracy [%]	
		OC Model	SVM$_{lin}$
Segmentation task	192	60	60

6.5 DISCUSSION

6.5.1 BIOMIMETIC VERSUS CLASSICAL MACHINE OLFACTION APPROACHES TO THE PROBLEM OF ODOR PERCEPTION

For classification and segmentation, the biomimetic approach and the classical machine learning techniques have performed comparably when applied to chemosensory polymer data. For classification using the synthetic data, the same trend can be observed given a large enough cortex (10,000 units). The classical approaches have displayed superior performance when compared to the model with a smaller cortex size (2000 units). For segmentation using the synthetic data, the SVM-based approach has shown a tendency to deliver higher success rates than the biomimetic model, particularly for binary (sweeping concentrations tests) and ternary mixtures. However, robust SVM performance required meticulous adjustment of the probability threshold, particularly for the set consisting of middle-concentration trials only, and was accompanied by a higher number of false positive detections for a sweeping concentrations paradigm. Also, for all the segmentation experiments the OC model was set to a smaller size than for the corresponding classification tests.

This study has provided insight into odor recognition mechanisms in the biomimetic model and cast light on potential refinements of the model that allow for matching the performance of the classical approaches. It has also raised questions about biological plausibility of coding odor information in our OC model, and in consequence, it has enhanced our understanding of the biological structure of the olfactory system.

Inhibitory feedback from the OC to the OB was used to demonstrate context dependence in the same network in which the segmentation takes place. This has served in this work as a proof of concept that such a top-down filter is able to enhance sensitivity to a task-relevant component in a complex background. This example has also illustrated that biomimetic models can be used not only to solve tasks independently of each other, but also as integrative platforms that combine a range of capabilities to perform various tasks in one integrated system.

Experiments on rats have shown that the piriform cortical pattern recognition and segmentation capabilities are shaped by task demands (Chapuis and Wilson 2011); i.e., learning occurs not only to increase discrimination capabilities but also to ignore differences between input mixtures when required by changing their learned cortical odor representations. In a natural setting this type of capability seems of great importance, as some tasks may benefit from increased discrimination capability, while others may benefit from increased generalization. The scenarios we have considered here, despite some biomimetic features, are well suited for classical machine olfaction problems. Therefore, the performance reported in this study may not necessarily be representative for future, more naturalistic real-world demands. More complex scenarios involving a combination of classical odor recognition tasks, such as classification and segmentation, appear as a next suitable evaluation step. In those cases that could additionally have to integrate input data from more than one type of sensor in changing environments and adjustable demands, the generality and flexibility of biomimetic models may be necessary.

6.5.2 BIOLOGICAL RELEVANCE

As emphasized earlier, the presented model has been designed with the aim of studying computational principles of olfactory information processing in mammals. The level of abstraction has been determined as a compromise between biological fidelity of the neural computational units in the network and the need to capture information-relevant components of population responses. As one of the key consequences, the focus has been on spatial aspects of olfactory representations at every processing stage, and their time-domain characteristics have not been accounted for. Despite the abundance of studies investigating temporal phenomena in an olfactory system, such as spike synchronization (Mori et al. 1999), stimulus-locked firing patterns (Christensen et al. 1998), oscillatory activity of local field potentials, and their coherence (Kay et al. 2009), a consensual view is emerging that the information about odor identity and intensity should be reliably, though perhaps not exclusively, contained in static components of the response (Huerta et al. 2004; Olsen et al. 2010). Therefore, despite the limited fidelity of the model in mimicking biological processes, the potential to generate new insights into principles of olfactory computations with distributed and stationary rate-based representations of odor stimuli is not really diminished. Crucially, the proposed level of abstraction that relies on simplistic computational units and disregards biophysical aspects of neural information transmission does not hinder the analysis of the dynamics of the attractor network, which underlies the simulated odor recognition function.

The model of the local ORN responses to a simulated olfactory stimulus has been largely inspired by Rospar et al.'s (2000) theoretical work on concentration-dependent activation characteristics of individual ORNs and Grosmaitre et al.'s (2006) experimental results demonstrating high variability in the activity patterns of

ORNs within the same OR group. The population response of the entire ORN array depended on the distribution of odor-OR affinity constants, which was generated with the objective to account for available biological data, and hence to reproduce the major statistical features of biological olfactory codes (Hopfield 1999; Malnic et al. 1999). In the computational literature, other approaches to synthesizing olfactory input to the OB have received more attention. A representative example is simulating so-called virtual response patterns (Schmuker and Schneider 2007). The authors embedded the chemical structure of real odorants in a high-dimensional physicochemical space and applied a self-organizing map to model the data distribution. The units of the topographically organized map served then as virtual sensors responding to given odors depending on the distances in the projected metric space. Compared to our approach, Schmuker and Schneider (2007) operated in lower-dimensional odor domains and strongly relied on the underlying topographical organization of the physicochemical odor features.

The novel finding about the heterogeneity of stimulus intensity responses of ORNs converging onto the same glomeruli (Grosmaitre et al. 2006) has paved the way for the proposed hypothesis of interval concentration coding at the next processing stage in the OB (Sandström et al. 2009). The biological underpinning of the emergent secondary olfactory representation lies in a competition between MT cells mediated by periglomerular and granule cell inhibition within the glomerular module. This concept goes beyond traditional interpretations of glomerular computations. According to one of the conventional views, glomeruli play only a role in sites where ORNs converge, and the above-mentioned signal-to-noise enhancement, along with nonspecific gain (Poo and Isaacson 2009), constitute their only functional principle (Chen and Shepherd 2005). However, there has been a growing body of experimental evidence for richer functionality of the glomerular circuitry in the OB in vertebrates, which has been manifested in the stimulus coding represented by MT cells (Wachowiak et al. 2002). It has been observed that high odorant concentrations lead to broader representations and monotonically greater activity among ORNs but not among MT cells (Cleland 2010). In other studies, strong inhibitory effects discovered particularly in the OB activity have been discussed in the context of local mechanisms of lateral inhibition aimed at sharpening individual neuron's tuning curves. This view has been widely debated, and consequently, other hypotheses as to the functional role of the inhibitory circuitry in the OB have been formulated (Cleland et al. 2006). One of the most prevailing concepts has been that of global normalization of the activity in the OB in the presence of varying stimulus intensity levels. Such normalization would enable relational concentration-invariant representations considering the relative pattern of activity among MT cells as the best predictor of odor identity (Cleland et al. 2007).

The cortical population is set up with a modular structure (Douglas et al. 1995), and each module has a winner-takes-all operation where only one unit will be active. This results in a binary, sparse, and distributed code that is suitable for associative storage of odor memories. Experiments show sparse and highly distributed odor representations in the piriform cortex (Isaacson 2010).

6.5.3 ODOR RECOGNITION BASED ON CONCENTRATION-DEPENDENT EARLY OLFACTORY RESPONSES

One of the fundamental questions underlying our understanding of the sense of smell is concerned with the representation of key stimulus features. In this regard it still remains unclear how the olfactory system conveys the information about odor intensity to facilitate robust recognition in natural environments. It has been demonstrated on numerous occasions that the information about the concentration of a source odor is preserved to a certain degree at each early olfactory processing stage (Cleland 2010) and in the stimulus representations in the piriform cortex (Stettler and Axel 2009). The principle of concentration coding has been most effectively explored at the level of ORNs (Buck 2006; Rubin and Katz 1999; Rospars et al. 2000, 2003; Lansky and Rospars 1993). The sigmoidal dose-response curves used in this study to characterize the ORN's behavior result in so-called concentration-saturating primary representations and imply a growing number of activated glomeruli with increasing odor intensity until the saturation level is reached. This effect can be readily observed in our model. Still, it should be noted that the aggregated input to each glomerulus is less sensitive to concentration changes than the corresponding individual ORN responses (Cleland 2010; Friedrich and Korsching 1997). The picture of stimulus intensity coding becomes far more obscure when studying the secondary olfactory representation at the output of the OB in the MT layer. As mentioned earlier in the discussion, a plethora of hypothetical views on computational aspects of OB neural circuits have been proposed with the assumption that the resulting distributed activation patterns are considerably less concentration dependent than the primary ORN codes (Cleland et al. 2007, 2011; Wachowiak et al. 2002). Although some evidence of concentration-invariant neural responses of the MT or the corresponding projection neurons in insects has been reported (Stopfer et al. 2003), this issue still remains a subject of intensive debate (Cleland et al. 2011). Our proposed interval concentration coding scheme conveys a novel concept of how the information about stimulus intensity could be handled in the mammalian olfactory system. As a result, the MT code is more sensitive to varying odor concentrations than the ORN representation, which is manifested in relatively larger distances (Euclidean or correlation-based metric) between the activity patterns induced at the output of the OB by the same odor applied at different intensity levels. The increased intra- versus inter-odor class variance for the *mtF* data set compared to the *ornF* had a noticeable effect on the concentration-dependent characteristics of olfactory pattern recognition. The fact that MT patterns are more decorrelated and sparse, with lower activity levels, than their ORN counterparts complies with the decorrelating function of the OB reported based on experimental studies (Friedrich and Laurent 2004).

The ecological relevance of the mammals' capability to detect different levels of intensity of olfactory stimuli has been discussed in depth (e.g., Gottfried 2009). Similarly, the relation between odor concentration and its perceived quality and strength has been the subject of extensive psychophysical studies (Furudono et al. 2009). However, regardless of how effective the olfactory system is or should be at perceptual discrimination of odor concentrations, its capacity to recognize the identity of the odor stimulus independently of its strength appears as a highly desirable feature—in natural environments chemical signals are experienced at varying, often quickly fluctuating, concentration levels (Riffell et al. 2009). This fundamental form of perceptual invariance has been widely reported in the olfactory domain (Gottfried 2009). Yet, the question as to how odors maintain their perceptual characteristics over a wide concentration range has not received an unequivocal and convincing answer. Concentration-invariant coding of olfactory stimuli has been proposed to take place in the OB, and a few alternative OB computational mechanisms have been suggested (Uchida and Mainen 2007; Cleland et al. 2007; Cleland and Sethupathy 2006; Chen and Shepherd 2005). Our hypothetical explanation of this perceptual generalization phenomenon rests on the principle of learned invariance, studied mostly in the visual modality (Caponnetto et al. 2008). It implies that the system can learn to associate object instances with some level of variation with one target category, provided that it has been exposed to a suitably representative training sample. This concept has attracted rather limited attention in the olfaction research community to date (Cleland et al. 2011; Uchida and Mainen 2007). We argue for the plausibility of this approach to concentration-independent recognition on the grounds of broad availability of olfactory stimuli at widely varying intensity levels in natural environments (Gottfried 2009; Hopfield 1999), as emphasized earlier. The results we obtained in odor classification and segmentation scenarios suggest a considerable level of generalization, still dependent on the learning machine. In this regard, supervised methods are favored due to their inherent mechanism of binding feature representations to class labels.

6.5.4 CODING, RECOGNITION, AND PERCEPTION OF ODOR MIXTURES

Odorants encountered in natural environments are dominated by multiligand mixtures and their blends. They are often composed of a large and complex variety of chemical components. Some of the key questions in the olfactory research to date have revolved around the problem of processing odor mixtures in the olfactory systems—and ultimately their perception. A considerable body of experimental evidence reveals a tendency for synergistic effects, and thus implies that the sense of smell is synthetic in nature; i.e., perception of mixtures is often generalized and categorized into discrete entities (Livermore and Laing 1998), usually resulting in the emergence of novel perceptual qualities that were not present in each component (Wiltrout et al. 2003). Under these circumstances, it appears that the capacity to identify single components in complex blends is rather limited (Jinks and Laing 1999), which entails some form of interaction of their representations in the olfactory pathway. This phenomenon has been traditionally attributed either to the OR binding processes (Rospars et al. 2008; Kay et al. 2003) or OB computations (Tabor et al. 2004). The interactions occur in a wide spectrum of different forms, depending mainly on the receptor characteristics and the combinations of odor molecules (Rospars et al. 2008; Kay et al. 2003). Among the most common categories are mixture suppression and enhancement (Rospars et al. 2008). In our computational study, the process of mixing odor molecules is modeled at the OR level, and it complies with one of the most predominant hypoadditive syntopic interaction types (Rospars et al. 2008). It produces a response to a mixture stimulus that is roughly equivalent to or slightly higher than the response to the most sensitive component. In cases of ligands with similar OR binding characteristics, the activity of the corresponding ORNs induced by their blend can be more synergistic in nature. Further processing of odor mixture representations in our model, both in the OB and in the OC, is then just a direct consequence of early ORN coding.

Despite the predominance of configural percepts of mixtures in biological olfaction, numerous psychophysical studies have demonstrated a certain level of capability of humans and rodents to segment complex odorants into chemically simpler elements (Miyazawa et al. 2009; Marshall et al. 2006; Laing and Francis 1989). These findings, along with a growing body of electrophysiological and neuroimaging evidence, suggest that at least in some cases the information about individual components must persist in the olfactory pathway; i.e., their representations do not undergo any significant interactions. The urgent question to be addressed in the olfactory research on mixture perception relates to key factors that determine the type of mixture processing and encoding in the system—elemental or configural. Although no unifying view has emerged yet, there have been a number of increasingly convincing hypotheses proposed based on experimental studies. It has been suggested that the degree of generalization strongly depends on the components' qualities, specifically on their perceptual similarity (Wiltrout et al. 2003). According to this concept, simple odors are more likely to be processed analytically when mixed if their individual percepts are distinguishable. This hypothesis has been extended to the similarity in the space of OB activation patterns induced by elemental odors. However, the so-called glomerular overlap hypothesis of mixture perception (cf. Frederick et al. 2009) has not really received enough support in biological data. Familiarity and exposure to components in isolation have also been seen as the basis for the type of processing of odor mixtures (Livermore and Laing 1998; Rabin and Cain 1989). In the same spirit, the ecological characteristics of odorants play an important role (Gottfried 2009). Finally, there has recently been convincing evidence gathered to support the view that the potential for elemental perception of some mixtures can be exploited under specific demands of a behavioral task the subject is engaged in (Wiltrout et al. 2003). Some of the proposed explanations point toward the significance of learning and association phenomena ascribed to the cortical function, hence implying the critical role of so-called olfactory perceptual learning (Gottfried 2010; Wilson and Stevenson 2003). Overall, it appears that both analytic and synthetic mechanisms are involved in coding and processing of single and complex odors to serve the overriding objective of robust recognition of environmentally and behaviorally relevant olfactory cues (Livermore and Laing 1998). It should be emphasized at this point that our model in the present form does not account for any explicit OC-related factors (indicative of cortical perceptual learning) determining the type of mixture processing. The outcome depends on the similarity of ORN and OB representations of individual odor components in agreement with the aforementioned glomerular overlap hypothesis of mixture perception (Frederick et al. 2009). Similarly, the effect of impeded segmentation capability at low concentrations (Duchamp-Viret et al. 1990) is reproduced.

6.5.5 Biological Underpinning and Models
of Odor Mixture Segmentation

In our OC model, recurrent connections were set up to span almost the entire piriform cortex, similarly to experimental reports (Wilson and Sullivan 2011). These connections have two functional roles. First, they make the cortex into a content-addressable memory (Haberly 1985; Haberly and Bower 1989). That is, by providing a part or a noisy version of an odor, the full memorized odor is retrieved. This so-called pattern completion has been reported to take place in the anterior piriform cortex (Barnes et al. 2008; Wilson 2009). Second, they are together with adaptation underlying the proposed segmentation mechanism. Our model differs from proposed segmentation mechanisms that utilize inhibitory feedback from the piriform cortex to the OB (Zhaoping and Hertz 2000; Raman and Gutierrez-Osuna 2005). There, classified odors can be successively removed from the OB representation of the mixture. Instead, we leave the OB representation unchanged during segmentation consistent with data from Kadohisa and Wilson (2006). The segmentation takes place by switching between odor representations in the cortical population by utilizing an adaptation that depends on the current state. Recordings from the anterior piriform cortex show that adaptation is highly odor specific (Wilson 2000). In addition, behavioral data show that short-term odor habituation memory (on a timescale of tens of seconds) can be blocked by the same pharmacological agents as synaptic adaptation (Linster et al. 2009, who also modeled this habituation).

For the adaptation in the segmentation tasks we used unit adaptation for the cortical population, which will be largely equivalent to synaptic adaptation if the odor representations are nonoverlapping in the cortical population. This was the case in the segmentation experiments performed. Contrary to models for background suppression (in OB; Gutiérrez-Gálvez and Gutierrez-Osuna 2006) and habituation (between OB and piriform cortex pyramidal cells; Linster et al. 2009) that also use adaptation mechanisms, we propose that odor segmentation takes place by adaptation in the recurrent connections in the cortical population. This could be on a faster timescale than the seconds to minutes that have been reported for homosynaptic depression of mitral cell input to the anterior piriform cortex (Best and Wilson 2004; Kadohisa and Wilson 2006) and used in models as a mechanism for odor background segmentation (Linster et al. 2007). Mechanisms with fast timescales are necessary as psychophysical data show a limit of around 200 to 300 ms for many perceptual tasks (Uchida et al. 2006). For example, it has been seen that for mixtures of two odors, rats only need to take one sniff (less than 200 ms) to make a discrimination decision of maximum accuracy (Uchida and Mainen 2003). Also, in human experiments on identification of components in binary mixtures, 50% were correctly found upon 1.5 s exposure to a mixture (Jinks and Laing 1999), which, if the identification of both mixtures utilizes segmentation, suggests a faster mechanism than the aforementioned homosynaptic depression of mitral cell input to the anterior piriform cortex. Mechanisms underlying cortical adaptation used in recurrent networks, such as synaptic depression, can be expressed on a below 1 s timescale (Tsodyks et al. 2000; Lundqvist et al. 2006). Possibly, a variation of adaptation timescales may suit different perceptual needs.

6.6 SUMMARY AND CONCLUSIONS

The objective of this chapter has been to compare a novel biomimetic model approach with machine learning methods on odor recognition. The synthetic olfactory data with biologically plausible characteristics and polymer sensor recordings have served as input. The problems of concentration-dependent identification and the mixture processing in the olfactory system have been addressed. Specifically, we have proposed an interval coding mechanism for the OB and a cortical adaptation-based mechanism for segmentation, and have demonstrated their use in different mixing paradigms. Context-dependent segmentation by inhibitory connections between the OC and the OB has also been shown. We have evaluated the holistic processing capabilities in the machine olfaction problems of classification and segmentation. The model has produced results comparable with those of state-of-the-art machine learning solutions on the polymer sensor data and in the classification task on the synthetic data provided that a large cortical population is used. It has revealed a lower level of performance in the most difficult tasks for the biomimetic model with a smaller number of cortical units on the synthetic data, which qualitatively match biological data and can lead to new insights on biological odor coding. Optimal performance does not seem critical since biological systems may not necessarily be optimized toward discrimination in specific scenarios like those studied here. These tasks, despite biomimetic features, may not have particularly strong biological relevance. Other scenarios involving a lower degree of supervision and a lower number of labeled training data should be investigated. Importantly, the biomimetic model has a capability of integrating several mechanisms in one system, which has been demonstrated in the task combining segmentation and context dependence. This type of feature may prove necessary when devising systems aimed at performing more complex tasks.

The evaluation of the polymer sensor data has also revealed potential for the practical applicability of the biomimetic approach to chemosensory object recognition problems. Despite the exploratory nature of this evaluation, we envisage the future relevance of the biomimetic methodology in addressing complex real-world chemosensory tasks in novel situations and applications.

ACKNOWLEDGMENTS

We thank Benjamin Auffarth for the preprocessing and characterization of the polymer data. This work was supported by StratNeuro (hosted by Karolinska Institutet) and EU projects NEUROCHEM (FP7 FET Project 216916) and BrainScaleS (FP7 FET Integrated Project 269921).

REFERENCES

Barnes, D. C., Hofacer, R. D., Zaman, A. R., Rennaker, R. L., and Wilson, D. A. 2008. Olfactory perceptual stability and discrimination. *Nat. Neurosci.* 11(12): 1378–1380.

Benjaminsson, S., and Lansner, A. 2011. Extreme scaling of brain simulations. In Jülich Blue Gene/P Extreme Scaling Workshop 2011, ed. B. Mohr and W. Fring. Technical Report FZJ-JSC-IB-2011-02, 7-10, Forschungszentrum, Jülich.

Best, A. R., and Wilson, D. A. 2004. Coordinate synaptic mechanisms contributing to olfactory cortical adaptation. *J. Neurosci.* 24(3): 652–660.

Bezdek, J. C. 1981. *Pattern recognition with fuzzy objective function algorithms.* New York: Plenum.

Buck, L. B. 1996. Information coding in the vertebrate olfactory system. *Annu. Rev. Neurosci.* 19: 517–544.

Caponnetto, A., Poggio, T., and Smale, S. 2008. *On a model of visual cortex: Learning invariance and selectivity from image sequences. CBCL/CSAIL Technical Report 2008-030. Cambridge, MA: MIT.*

Chapuis, J., and Wilson, D. A. 2011. Bidirectional plasticity of cortical pattern recognition and behavioral sensory acuity. *Nat. Neurosci.* 15(1): 155–161.

Chen, W. R., and Shepherd, G. M. 2005. The olfactory glomerulus: A cortical module with specific functions. *J. Neurocytol.* 34: 353–360.

Christensen, T. A., Waldrop, B. R., and Hildebrand, J. G. 1998. Multitasking in the olfactory system: Context-dependent responses to odors reveal dual GABA-regulated coding mechanisms in single olfactory projection neurons. *J. Neurosci.* 18: 5999–6008.

Cleland, T. A. 2010. Early transformations in odor representation. *Trends Neurosci.* 33(3): 130–139.

Cleland, T. A., Chen, S.-Y. T., Hozer, K. W., Ukatu, H. N., Wong, K., and Zheng, F. 2011. Sequential mechanisms underlying concentration invariance in biological olfaction. *Frontiers Neuroeng.* 16(4): 21.

Cleland, T. A., Johnson, B. A., Leon, M., and Linster, C. 2007. Relational representation in the olfactory system. *Proc. Natl. Acad. Sci. U.S.A.* 6: 1953–1958.

Cleland, T. A., and Linster, C. 2005. Computation in the olfactory system. *Chem. Senses* 30: 801–813.

Cleland, T. A., and Sethupathy, P. 2006. Non-topographical contrast enhancement in the olfactory bulb. *BMC Neurosci.* 7.

Cover, T. M., and Hart, P. E. 1967. Nearest neighbor pattern classification. *IEEE Trans. Information Theory* 13(1): 21–27.

Distante, C., Ancona, N., and Siciliano, P. 2003. Support vector machines for olfactory signals recognition. *Sensors Actuators B* 88: 30–39.

Douglas, R. J., Koch, C., Mahowald, M., Martin, K. A., and Suarez, H. H. 1995. Recurrent excitation in neocortical circuits. *Science* 269(5226): 981–985.

Duchamp-Viret, P., Duchamp, A., and Sicard, G. 1990. Olfactory discrimination over a wide concentration range.Comparison of receptor cell and bulb neuron abilities. *Brain Res.* 517: 256–262.

Frederick, D. E., Barlas, L., Ievins, A., and Kay, L. M. 2009. A critical test of the overlap hypothesis for odor mixture perception. *Behav. Neurosci.* 123: 430–437.

Friedrich, R. W., and Korsching, S. I. 1997. Combinatorial and chemotopic odorant coding in the zebrafish olfactory bulb visualized by optical imaging. *Neuron* 18: 737–752.

Friedrich, R. W., and Laurent, G. 2004. Dynamics of olfactory bulb input and output activity during odor stimulation in zebrafish. *J. Neurophysiol.* 91: 2658–2669.

Furudono, Y., Yukio, S., Kayori, T., Junzo, H., and Takaaki, S. 2009. Relationship between peripheral receptor code and perceived odor quality. *Chem. Senses* 34: 151–158.

Gottfried, J. A. 2009. Function follows from ecological constraints on odor codes and olfactory percepts. *Curr. Opin. Neurobiol.* 19: 422–429.

Gottfried, J. A. 2010. Central mechanisms of odour object perception. *Nat. Rev. Neurosci.* 11(9): 628–641.

Grosmaitre, X., Vassalli, A., Mombaerts, P., Shepherd, G. M., and Ma, M. 2006. Odourant responses of olfactory sensory neurons expressing the odourant receptor MOR23: A patch clamp analysis in gene-targeted mice. *Proc. Natl. Acad. Sci. U.S.A.* 103: 1970–1975.

Gutiérrez-Gálvez, A., and Gutierrez-Osuna, R. 2006. Contrast enhancement and background suppression of chemosensor array patterns with the KIII model. *Int. J. Intell. Syst.* 21(9): 937–953.

Gutierrez-Osuna, R. 2002. Pattern analysis for machine olfaction: A review. *IEEE Sensors J.* 2: 189–202

Haberly, L. B. 1985. Neuronal circuitry in olfactory cortex: Anatomy and functional implications. *Chem. Senses* 10(2): 219–238.

Haberly, L. B., and Bower, J. M. 1989. Olfactory cortex: Model circuit for study of associative memory? *Trends Neurosci.* 12(7): 258–264.

Hines, E. L., Boilot, P., Gardner, J. W., and Gongora, M. A. 2003. Pattern analysis for electronic noses. In *Handbook of machine olfaction: Electronic nose technology*, ed. T. C. Pearce, S. S. Schiffman, H. T. Nagle, and J. W. Gardner, 133–160. Weinheim: Wiley.

Herman, P. A., and Lansner, A. 2010. Odour recognition framework for evaluating olfactory codes. In *Proceedings of ECRO XXI Congress*, Avignon, 2010.

Hopfield, J. J. 1999. Odor space and olfactory processing: Collective algorithms and neural implementation. *Proc. Natl. Acad. Sci. U.S.A.* 96: 12506–12511.

Huang, T.-K., Weng, R. C., and Lin C. J. 2006. Generalized Bradley-Terry models and multi-class probability estimates. *J. Machine Learning* 7: 85–115.

Huerta, R., Nowotny, T., Garcia-Sanchez, M., Abarbanel, H. D. I., and Rabinovich, M. I. 2004. Learning classification in the olfactory system of insects. *Neural Comput.* 16: 1601–1640.

Isaacson, J. 2010. Odour representations in mammalian cortical circuits. *Curr. Opin. Neurobiol.* 20(3): 328–331.

Jinks, A., and Laing, D. G. 1999. Temporal processing reveals a mechanism for limiting the capacity of humans to analyze odor mixtures. *Cognitive Brain Res.* 8(3): 311–325.

Kadohisa, M., and Wilson, D. A. 2006. Olfactory cortical adaptation facilitates detection of odors against background. *J. Neurophysiol.* 95(3): 1888–1896.

Kay, L. M., Beshel, J., Brea, J., Martin, C., Rojas-Líbano, D., and Kopell, N. 2009. Olfactory oscillations: The what, how and what for. *Trends Neurosci.* 32: 207–214.

Kay, L. M., Lowry, C. A., Jacobs, H. A. 2003. Receptor contributions to configural and elemental odor mixture perception. *Behav. Neurosci.* 117: 1108–1114.

Kraskov, A., and Grassberger, P. 2009. MIC: Mutual information based hierarchical clustering. In *Information theory and statistical learning*, 101–123. New York: Springer.

Laing, D., and Francis, G. 1989. The capacity of humans to identify odors in mixtures. *Physiol. Behav.* 46(5): 809–814.

Lansky, P., and Rospars, J.-P. 1993. Coding of odour intensity. *Biosystems* 31: 15–38.

Lansner, A., Benjaminsson, S., and Johansson, C. 2009. From ANN to biomimetic information processing. In *Biologically inspired signal processing for chemical sensing*, ed. G. Agustín and M. Santiago, 33–43. Vol. 188. Heidelberg: Springer.

Lansner, A., and Holst, A. 1996. A higher order Bayesian neural network with spiking units. *Int. J. Neural Syst.* 7: 115–128.

Laurent, G. 1999. A systems perspective on early olfactory coding. *Science* 286: 723–728.

Linster, C., Henry, L., Kadohisa, M., and Wilson, D. A. 2007. Synaptic adaptation and odor-background segmentation. *Neurobiol. Learning Memory* 87(3): 352–360.

Linster, C., Menon, A. V., Singh, C. Y., and Wilson, D. A. 2009. Odor-specific habituation arises from interaction of afferent synaptic adaptation and intrinsic synaptic potentiation in olfactory cortex. *Learning Memory* 16(7): 452–459.

Livermore, A., and Laing, D. G. 1998. The influence of chemical complexity on the perception of multicomponent odor mixtures. *Percept. Psychophys.* 60: 650–661.

Lundqvist, M., Rehn, M., Djurfeldt, M., and Lansner, A. 2006. Attractor dynamics in a modular network model of neocortex. *Network* 17(3): 253–276.

Malnic, B., Hirono, J., Sato, T., and Buck, L. 1999. Combinatorial receptor codes for odors. *Cell* 96: 713–723.

Marshall, K., Laing, D. G., Jinks, A. L., and Hutchinson, I. 2006. The capacity of humans to identify components in complex odor–taste mixtures. *Chem. Senses* 31: 539–545.

Miyazawa, T., Gallagher, M., Preti, G., and Wise, P. M. 2009. Psychometric functions for ternary odor mixtures and their unmixed components. *Chem. Senses* 34: 753–761.

Mori, Y., Nahao, H., and Yoshihara, Y. 1999. The olfactory bulb: Coding and processing odor molecule information. *Science* 286: 711–715.

Muniz, J. E., Cain, W. S., and Abraham, M. H. 2003. Dose-addition of individual odorants in the odor detection of binary mixtures. *Behav. Brain Res.* 138: 95–105.

Olsen, S. R., Bhandawat, V., and Wilson, R. I. 2010. Divisive normalization in olfactory population codes. *Neuron* 66: 287–299.

Pearce, T. C. 1997. Computational parallels between the biological olfactory pathway and its analogue "the electronic nose." Part I. Biological olfaction. *Biosystems* 41: 43–67.

Platt, J. 2000. Probabilistic outputs for support vector machines and comparison to regularized likelihood methods. In *Advances in large margin classifiers*, ed. A. Smola, P. Bartlett, B. Schölkopf, and D. Schuurmans, 61–74. Cambridge, MA: MIT Press.

Poo, C., and Isaacson, J. S. 2009. Odor representations in olfactory cortex: "Sparse" coding, global inhibition, and oscillations. *Neuron* 62: 850–861.

Rabin, M. D., and Cain, W. S. 1989. Attention and learning in the perception of odor mixtures. In *Perception of complex smells and tastes*, ed. D. G. Laing, W. S. Cain, R. L. McBride, and B. W. Ache, 173–178. Sydney: Academic Press.

Raman, B., and Gutierrez-Osuna, R. 2005. Mixture segmentation and background suppression in chemosensor arrays with a model of olfactory bulb-cortex interaction. In *Proceedings of IEEE International Joint Conference on Neural Networks 2005 (IJCNN '05)*, Montreal, Canada, 1: 131–136.

Riffell, J. A., Lei, H., Christensen, T. A., Hildebrand, J. G. 2009. Characterization and coding of behaviorally significant odor mixtures. *Curr. Biol.* 19: 335–340.

Rospars, J.-P., Lánský, P., Chaput, M., Duchamp-Viret, P., and Duchamp, A. 2008. Competitive and noncompetitive odorant interactions in the early neural coding of odorant mixtures. *J. Neurosci.* 28: 2659–2666.

Rospars, J.-P., Lánský, P., Duchamp, A., and Duchamp-Viret, P. 2003. Relation between stimulus and response in frog olfactory receptor neurons in vivo. *Eur. J. Neurosci.* 18: 1135–1154.

Rospars, J.-P., Lánský, P., Duchamp-Viret, P., and Duchamp, A. 2000. Spiking frequency versus odourant concentration in olfactory receptor neurons. *Biosystems* 58: 133–141.

Royet, J. P., Distel, H., Hudson, R., and Gervais, R. 1998. The number of glomeruli and mitral cells in the olfactory bulb of rabbit. *Brain Res.* 788: 35–42.

Rubin, B. D., and Katz, L. C. 1999. Optical imaging of odorant representations in the mammalian olfactory bulb. *Neuron* 23: 499–511.

Sandberg, A., Lansner, A., Petersson, K. M., and Ekeberg, O. 2000. A palimpsest memory based on an incremental Bayesian learning rule. *Neurocomputing* 32–33(1–4): 987–994.

Sandström, M., Proschinger, T., and Lansner, A. 2009. A bulb model implementing fuzzy coding of odor concentration. In *Proceedings of the 13th International Symposium on Olfaction and Electronic Nose*, Brescia, Italy, 1137: 159–162.

Schmuker, M., and Schneider, G. 2007. Processing and classification of chemical data inspired by insect olfaction. *Proc. Natl. Acad. Sci. U.S.A.* 104: 20285–20289.

Schölkopf, B., Burges, C. J. C., and Smola, A. J. 1998. *Advances in kernel methods—Support vector learning*. Cambridge, MA: MIT Press.

Stettler, D. D., and Axel, R. 2009. Representations of odor in the piriform cortex. *Neuron* 63: 854–864.

Stopfer, M., Jayaraman, V., and Laurent, G. 2003. Intensity versus identity coding in an olfactory system. *Neuron* 39: 991–1004.

Tabor, R., Yaksi, E., Weislogel, J. M., and Friedrich, R. W. 2004. Processing of odor mixtures in the zebrafish olfactory bulb. *J. Neurosci.* 24: 6611–6620.

Tsodyks, M. V., Uziel, A., and Markram, H. 2000. Synchrony generation in recurrent networks with frequency-dependent synapses. *J. Neurosci.* 20(1): RC50.

Uchida, N., Kepecs, A., and Mainen, Z. F. 2006. Seeing at a glance, smelling in a whiff: Rapid forms of perceptual decision making. *Nat. Rev. Neurosci.* 7(6): 485–491.

Uchida, N., and Mainen, Z. F. 2003. Speed and accuracy of olfactory discrimination in the rat. *Nat. Neurosci.* 6(11): 1224–1229.

Uchida, N., and Mainen, Z. F. 2007. Odor concentration invariance by chemical ratio coding. *Front. Syst. Neurosci.* 1: 3.

Ueda, N., and Nakano, R. 1994. A new competitive learning approach based on an equidistortion principle for designing optimal vector quantizers. *Neural Networks* 7(8): 1211–1227.

Wachowiak, M., Cohen, L. B., Zochowski, M. R. 2002. Distributed and concentration-invariant spatial representations of odorants by receptor neuron input to the turtle olfactory bulb. *J. Neurophysiol.* 87: 1035–1045.

Wilson, D. A. 2000. Comparison of odor receptive field plasticity in the rat olfactory bulb and anterior piriform cortex. *J. Neurophysiol.* 84(6): 3036–3042.

Wilson, D. A. 2009. Pattern separation and completion in olfaction. *Ann. N.Y. Acad. Sci.* 1170(1): 306–312.

Wilson, D. A., and Stevenson, R. J. 2003. Olfactory perceptual learning: The critical role of memory in odor discrimination. *Neurosci. Biobehav. Rev.* 27: 307–328.

Wilson, D. A., and Sullivan, R. M. 2011. Cortical processing of odor objects. *Neuron* 72 (4): 506–519.

Wiltrout, C., Dogra, S., and Linster, C. 2003. Configurational and nonconfigurational interactions between odorants in binary mixtures. *Behav. Neurosci.* 117: 236–245.

Winkler, R., Klawonn, F., and Kruse, R. 2011. Fuzzy C-means in high dimensional spaces. *Int. J. Fuzzy Syst. Appl.* 1: 1–16.

Wold, H. 1966. Estimation of principal components and related models by iterative least squares. In *Multivariate analysis*, ed. P. R. Krishnaiah, 391–420. New York: Academic Press.

Wold, H. 1985. Partial least squares. In *Encyclopedia of statistical sciences*, ed. S. Kotz and N. L. Johnson, 581–591. Vol. 6. New York: Wiley.

Yea, B., Konishi, R., Osaki, T., and Sugahara, K. 1994. The discrimination of many kinds of odour species using fuzzy reasoning and neural networks. *Sensors Actuators A* 45: 159–165.

Young, F. W. 1985. Multidimensional scaling. In *Encyclopedia of statistical sciences*, ed. S. Kotz, N. L. Johnson, and C. B. Read. Vol. 5. New York: Wiley.

Zhaoping, L., and Hertz, J. 2000. Odour recognition and segmentation by a model olfactory bulb and cortex. *Network* 11(1): 83–102.

Index

T - #0415 - 071024 - C8 - 234/156/11 - PB - 9780367380151 - Gloss Lamination